SpringerBriefs in Computer Science

Series Editors

Stan Zdonik
Peng Ning
Shashi Shekhar
Jonathan Katz
Xindong Wu
Lakhmi C. Jain
David Padua
Xuemin Shen
Borko Furht
V. S. Subrahmanian
Martial Hebert
Katsushi Ikeuchi
Bruno Siciliano

For further volumes:
http://www.springer.com/series/10028

Betsy George · Sangho Kim

Spatio-temporal Networks

Modeling and Algorithms

 Springer

Betsy George
Oracle Inc.
Nashua, NH
USA

Sangho Kim
Esri
Redlands, CA
USA

ISSN 2191-5768   ISSN 2191-5776 (electronic)
ISBN 978-1-4614-4917-1   ISBN 978-1-4614-4918-8   (eBook)
DOI 10.1007/978-1-4614-4918-8
Springer New York Heidelberg Dordrecht London

Library of Congress Control Number: 2012943367

Printed on acid-free paper

Springer is part of Springer Science+Business Media (www.springer.com)

To the loving memory of my grandfather for his unconditional love and relentless encouragement!!

Betsy George

To my daughter Ellie and wife Taeeun in appreciation of their patience and understanding

Sangho Kim

Preface

Spatio-temporal networks are spatial networks whose topology and/or attributes change with time. These are encountered in many critical areas of everyday life such as transportation networks, electric power distribution grids, and social networks of mobile users. With the advances in technology, monitoring the temporal changes of such networks is becoming increasingly easier. For example, the increasing use of traffic sensors on transportation networks generates large volumes of data such as congestion levels and it becomes important to incorporate these data into data models and algorithms that deal with spatio-temporal networks.

A spatio-temporal network (STN) typically consists a finite set of nodes with location attributes, relationships between nodes (aka edges), and time-dependent attributes associated with nodes and relationships. STN modeling and computations raise significant challenges. The model must meet the conflicting requirements of simplicity and adequate support for efficient algorithms. Another challenge is to address the change in semantics of common graph operations such as shortest path computation, when temporal dimension is added. For example, shortest path between a start and an end location might be different at different times of day. Also paradigms (e.g., dynamic programming) used in algorithm design may be ineffective since their assumptions (e.g., stationary ranking of candidates) may be violated by the dynamic nature of STNs.

In recent years, STNs have attracted considerable attention in reserach. New representations have been proposed along with algorithms to perform key STN operations, while accounting for their time dependence. Designing a STN database would require the development of data models, query languages, indexing methods to efficiently represent, query, store, and manage time-variant properties of the network.

This book explores this design at conceptual, logical, and physical level. Models used to represent STNs are explored and analyzed. STN operations with emphasis on their altered semantics with addition of temporal dimension, are addressed, illustrating the capability toward answering interesting questions. For example, it is possible to answer queries such as, When is the best time to start so that I spend the least time on the road? Algorithms to implement these network operations are discussed. A comparative study of various models and algorithms would also be provided.

Nashua, NH, USA, April 2012   Betsy George
                                                                     Sangho Kim

Acknowledgments

First, I would like to thank Prof. Shashi Shekhar, Professor, Department of Computer Science, University of Minnesota, my advisor, for all the support and guidance during my research career. Without his constant encouragement this book would not have been possible. Thank you Prof Shekhar, for being an amazing mentor. You inspire me with your remarkable and unmatched wisdom, intellect, and knowledge.

My heartfelt thanks go out to my manager Dr. Jack Wang and Huiling Gong of Network Data Model group at Oracle Corporation for making every conversation on Spatio-temporal networks, exciting and fruitful.

I would like to thank my amazing friends for their incredible support; your faith in me keeps me going. Susan, Amy, Anjali, Vishal, Jean, Erin; thank you for your amazing friendship, love, support, laughter, and fun!! I count you as the blessings of my life!!

Last, but not certainly the least to my family, I cannot begin to express how grateful I am for your love!!!

I thank my father, for being the support that he has been, unrelenting, reliable and solid. You gave me the courage to step out into the world and explore and taught me to be open to ideas, to respect everyone, and courage to stand firm on my own convictions; to my mother, for her joy and positive attitude even in the face of adversities, resilience, and her generous spirit. Your sacrifices and unconditional love have made me what I am today. Thank you, Biju, the best brother in the world, for being there for me, always. You are an incredible person and I am so blessed to be your sister!! Thank you to my uncles, Joseph Kurian and Paul Kurian, for being so generous with your life; without your steadfast love and support, I would not be where I am today.

Nashua, NH, USA   Betsy George

I have had the good fortune of being around many remarkable individuals who have helped me complete this book. First, I would like to thank my advisor, Professor Shashi Shekhar for his support and guidance as an incredible mentor.

I have truly appreciated the exceptional research training that you have provided me, the confidence you have instilled in me, and all the advice you have given me throughout my five years of working with you.

My thanks also go to my manager Dr. Erik Hoel of Geodatabase team at ESRI for providing me with valuable discussion and collaboration as well as giving constructive critiques about network model and Geodatabase. The completion of this book would not have been possible without the help of these individuals.

To my family—I love you with all my heart.

Last but certainly not least, to my love, my little family who gives meaning to my life, my beloved Taeeun and cutest Hayne (I still prefer her Korean name). I could not have done this without your love and support. Taeeun, thank you for treating my difficulties as if they were your own and being with me for finishing this book. Hayne, you did a good job by eating a lot, growing fast, and behaving well ever since you were born. Without you and your mom, nothing has meaning.

Redlands, CA, USA   Sangho Kim

Contents

Chapter 1
Spatio-temporal Networks: An Introduction

Abstract Spatio-temporal networks are spatial networks whose topology and parameters change with time. These networks are important for many critical applications such as emergency traffic planning and route finding services. This chapter establishes the significance of such networks by describing its applications and briefly outlines the basic concepts.

1.1 Spatio-temporal Networks

Spatio-temporal networks are encountered in many significant areas of everyday life, such as transportation, sensor networks, and crime analysis. The underlying data of interest in these domains display a network structure, where the connectivity between entities is as relevant as their locations and proximity. Also, a number of network attributes can display time dependence. A spatio-temporal network typically consists of a finite set of points with location information, relationships between pairs of points, and time dependent attributes attached to points and relationships. Static spatial networks have been traditionally represented as graphs, where nodes represented the points and edges modeled the relationships. This representation does not capture the temporal dimension of the networks and the computations based on this model could lead to inaccurate results. For example, computing the shortest routes in a transportation network without accounting for travel time changes due to varying levels of congestion during the day, might not give the correct results.

Figure 1.1 illustrates traffic sensor networks on urban highways which measure congestion levels at two different times (e.g. 5:07 and 9:37 p.m.). With the increasing use of sensor networks that monitor traffic data on spatial networks and the consequent availability of time-varying traffic data, it becomes important to incorporate this data into the models and algorithms related to transportation networks. However, existing spatio-temporal databases do not provide adequate support for spatio-temporal networks.

B. George and S. Kim, *Spatio-temporal Networks*,
SpringerBriefs in Computer Science, DOI: 10.1007/978-1-4614-4918-8_1,
© The Author(s) 2013

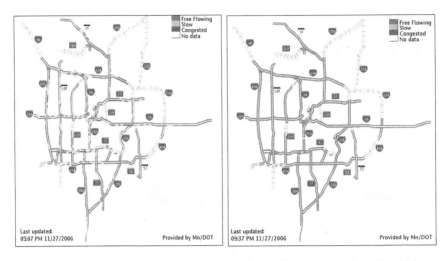

Fig. 1.1 Sensor networks periodically report time-variant traffic volumes on Twin Cities highways

Though commercial database systems such as Oracle (version 11g) [6, 7] and ArcMap from ESRI [13] provide support for spatial networks, they do not address the temporal aspects of the data. Existing work on graph databases [11, 29, 40, 44, 45] also do not adequately address time variance of spatial networks.

Designing a spatio-temporal network database would require the development of a new data model to efficiently store and manage time-variant properties of the network. The design of this new model would follow a three step process, namely conceptual, logical, and physical levels. In addition, new algorithms need to be formulated to process queries in specific application domains. These tasks raise significant computer science challenges such as (1) finding a balance between conflicting requirements of storage efficient and expressive power, (2) handling new and alternative semantics of graph operations, and (3) designing efficient and correct algorithms since some commonly assumed properties might not hold when temporal changes are considered.

The book will explore the design of spatio-temporal network databases at the conceptual, logical, and physical level. Existing approaches to incorporate temporal changes will be evaluated for their storage efficiency and support for computationally efficient algorithms.

1.2 Application Domain

An important application domain for spatio-temporal network databases is transportation science [21], a multi-disciplinary field that requires expertise from different domains. The difficulty, but also fascination, of this professional practice derives from the intrinsic complexity of transportation systems, which have both physical and behavioral elements.The physical elements in the systems (e.g., vehicles, infrastructure, etc.) are governed by the laws of physics. On the other hand, the mechanisms underlying the functionality and performance of these physical elements are often connected to travelers' behavioral choices. Traditionally the center of behavioral choice modeling [43] has been user equilibrium [48], the idea that all travelers use the least inconvenient routes and no individual can unilaterally improve his/her travel. A key assumption of user equilibrium is that travelers have perfect information about road conditions, and indeed this is generally true for commuters, who learn recurrent congestion patterns from their day-to-day travels. However, the assumption does not hold when the congestion is non-recurrent, in particular, when an extreme event occurs, and transportation network conditions become dynamic and uncertain. Thus one of the greatest challenges in transportation science is how to manage traffic in time-varying transportation networks, especially in disaster situations. This challenge cannot be met without the development of spatio-temporal databases. Currently, transportation management generates tremendous volumes of data and a large semantic gap exists between transportation science concepts and the concepts supported by current database systems. Emergency traffic management requires research in computer science to develop appropriate spatio-temporal database representations and query processing algorithms to make decisions in a timely manner.

Apart from emergency planning, spatio-temporal network modeling and planning have significant impact on applications such as daily commute routing and freight delivery services, where the primary focus is to reduce logistical costs such as fuel consumption. Commuters try to find a suitable time to start their commute (best start time) so that they spend the least time in traffic. The problem of finding best start time has similar applications in freight delivery services also. The importance of this problem was emphasized by the New York Times article reporting about a research by United Parcel Service of America Inc. (UPS). It said "The research at U.P.S. is paying off. Last year, it cut 28 million miles from truck routes saving roughly three million gallons of fuel in good part by mapping routes that minimize left turns" [9]. It is easy to recognize the spatio-temporal dimensions of this research since the shortest routes and most left turn restrictions are time dependent.

Another area where spatio-temporal routing finds application is in crime analysis. While mining the frequent routes traveled by criminals, it is necessary to consider the temporal nature of the transportation network (at times multimodal) to produce accurate results.

1.3 Background Information

Spatial Network Modeling: The purpose of conceptual modeling is to adequately represent the data types, their relationships and the associated constraints. Entity Relationship (ER) model, widely used in conceptual modeling, does not offer

Fig. 1.2 A PEER diagram for
a spatial network

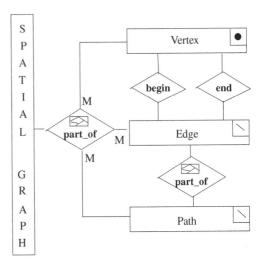

adequate features to capture the spatial semantics of networks. The most critical feature of spatial networks, namely the connectivity between objects can be expressed using a graph framework. At the conceptual level, the pictogram enhanced ER (PEER) model [46] can be used. Figure 1.2 shows a PEER diagram for a spatial network.

In the logical modeling phase, the conceptual data model is implemented using a commercial database management system. Spatial networks have been extensively modeled using graphs. In a spatial graph, vertices represent locations and edges represent relationships between locations. In a road network, the nodes could model intersections and edges the road segments connecting these intersections. Labels and weights can be attached to vertices and edges to encode additional information such as names and travel times. Two edges are considered to be adjacent if they share a common vertex.

The physical data modeling phase deals with the actual implementation of the database application. Issues related to storage, indexing and memory management are addressed in this phase. Adjacency list and adjacency matrix [5] are two well known main-memory data structures commonly used to implement graphs.

Time Expanded Graph: Time expanded graphs can be used to represent tim dependent graphs. Given a directed graph $G(V, E)$, we can define the time expanded graph G_T as follows.

Definition 1 (*Time Expanded Graph*) Let $G(N, E)$ be a directed network with set of node N and the set of edges E with travel time $\sigma_{i,j}$. The time expanded graph over a time horizon T is defined as:

$N_T := \{ N_{i,t} \mid i \in \text{node } id \text{ of } N \text{ and } t = 0, 1, \ldots, T \}$

$E_T := \{ (N_{i,t}, N_{i,t+1}) \mid i \in \text{node } id \text{ of } N \text{ and } t = 0, \ldots, \text{T-1} \}$

Fig. 1.3 Time expanded graph

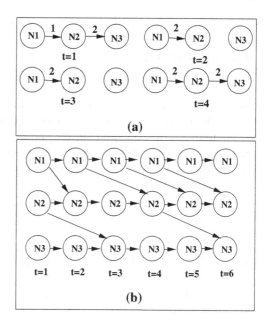

(a)

(b)

$$\cup \{(N_{i1,t1}, N_{i2,t2}) \mid i1, i2 \in \text{node } id \text{ of } N_{i1} \text{ and } N_{i2} \text{ and } t2 = t1 + \sigma_{i1,i2}\}$$

In the above definition, the $(N_{i,t}, N_{i,t+1})$ edge is called a holdover edge and the $(N_{i1,t1}, N_{i2,t2})$ edge is called a transfer edge. Figure 1.3b shows the time expanded graph for the snapshots of a network shown in Fig. 1.3a.

FIFO and non-FIFO network: In a network, if a flow arrives at the end node of each edge in the same order as they started at the start node of the edge, the edge travel costs are said to follow the First In First Out (FIFO) property and the edge is a FIFO edge. If every edge in a network follows FIFO properly, the network is a FIFO network. An edge that is not a FIFO edge is called a non-FIFO edge. If there is at least one non-FIFO edge in a network it is a non-FIFO network. Research in the area of spatio-temporal networks has primarily been conducted in the fields of databases and operations research. Related work in the field of databases fall into three broad categories (1) spatial network databases, (2) graph Databases, and (3) spatio-temporal databases. The recent release of Oracle (version 11g) includes a network data model to store and maintain the connectivity of link-node networks and supports basic features such as shortest path computation [33]. The Network Analyst extension of ArcMap from ESRI supports a network geodatabase and provides basic algorithms (e.g., shortest path, service area, closest facility, etc.) [13]. However, these products do not address the time variance of spatial networks, which is crucial in applications such as route computations and emergency planning. Although the need for live traffic information is increasing, there has been little work on the modeling and algorithms for spatio-temporal network databases. Chorochronos [27], studied various aspects of spatio-temporal databases including ontology, modeling, and implementation. However, researchers have yet to study spatio-temporal networks in this framework.

Graph databases [11–13, 40, 45, 47] primarily deal with spatial networks that do not vary with time. Research in graph databases that accounts for temporal variations perform computations over a snapshot of the network [10, 23, 38], and do not consider the interplay between the edge travel times and the existence of edges. Ding [10] proposed a model that addresses time-dependency by associating a temporal attribute to every edge and node of the network so that its state at any instant of time can be retrieved. This model performs path computations over a snapshot of the network. Since the network can change over the time taken to traverse these paths, this computation might not give realistic solutions. It does not propose an algorithm for the least travel time paths.

On conceptual level, various temporally enhanced entity relationship models have been proposed [19]. Some of these models capture the temporal properties of relationships in terms of their existence and validity periods; these do not explicitly capture the changes in relationship types. Other models such as TERC+ [50] capture the temporal nature of relationship types by expressing the relationship changes in terms of entity transformations. This model basically uses entity subtypes to represent temporal evolution of entities as well as relationships and hence might not be able to represent evolving relationships between entities without subtypes.

Research in Operations Research is based on the time expanded network [25, 26, 30, 36, 42]. This model duplicates the original network for each discrete time unit $t = 0, 1, \ldots, T$ where T represents the extent of the time horizon. The expanded network has edges connecting a node and its copy at the next instant in addition to the edges in the original network, replicated for every time instant. The approach significantly increases the network size and is very expensive with respect to memory. Because of the increased problem size due to replication of the network, the computations also become quite expensive. In addition, time expanded graphs have representational issues when modeling non-flow networks. Also, time expanded graphs require a prior knowledge of the length of the time period and hence might lead to a semantic mismatch while handling infinite time series. This model incorporates the time dependent edge attributes into the graph in the process of graph expansion making it more application- dependent, thus making physical data independence harder to achieve.

Stochastic models which use probability distribution functions to describe travel time [20, 25, 31, 32] have been used to study time-dependence of transportation networks. T hough they can give valuable insights into the traffic flow analysis, the computational cost to compute the least expected travel times in these networks is prohibitively large to adapt to real life scenarios [32].

Chapter 2
Time Aggregated Graph: A Model for Spatio-temporal Networks

Abstract Spatio-temporal networks represent networks where entities have spatial attributes and the topology and parameters display time-dependence. Given the significance of such networks in critical domains such as transportation science and sensor data analysis, the importance of a model that is simple, expressive and storage efficient to represent such networks cannot be understated. The model must provide support for the design of algorithms to process frequent queries that need to be answered in the application domains. This problem is challenging due to potentially conflicting requirements of model simplicity and support for efficient algorithms. Time expanded networks which have been used to model dynamic networks employ replication of the network across time instants, resulting in high storage overhead and algorithms that are computationally expensive. Time-aggregated graphs do not replicate nodes and edges across time; rather they allow the properties of edges and nodes to be modeled as a time series. The chapter presents a description and comparison of these models.

2.1 Modeling Spatio-temporal Networks

The growing importance of application domains such as transportation networks, emergency planning and location based services highlights the need for efficient modeling of spatio-temporal networks (e.g. road networks) that takes into account changes to the network over time. The model should provide the necessary framework for developing efficient algorithms that implement the frequent operations posed on such networks. Frequent queries on such networks might include finding the shortest route from one place to another or a search for the nearest neighbor. The shortest route would depend on the time dependent properties of the network such as congestion on certain road segments, which would increase the travel time on that segment. The result of a nearest neighbor search could also be time sensitive if it is based on a road network.

B. George and S. Kim, *Spatio-temporal Networks*,
SpringerBriefs in Computer Science, DOI: 10.1007/978-1-4614-4918-8_2,
© The Author(s) 2013

Modeling such a network poses many challenges. Not only should the model be able to accommodate changes and compute the results consistent with the existing conditions, it should do so accurately and simply. In addition, the need to answer frequent queries quickly means fast algorithms are required for computing the query results. The model should thus provide sufficient support for the design of correct and efficient algorithms for frequent computations.

Often dynamic networks have been modeled as time expanded networks, where the entire network is replicated for every time instant. The changes in the network, especially the travel time variations, can be very frequent and for modeling such frequent changes, the time expanded networks would require a large number of copies of the original network, thus leading to network sizes that are too memory expensive. For example, traffic sensors on highway networks send measurement data every 30 s. A one-year dataset may need over one million copies of the road network, which itself may have a million nodes and edges for each time instant. Such large sized networks would also result in computationally expensive algorithms.

Various temporally enhanced entity relationship models have been proposed [19]. Some of these models capture the temporal properties of relationships in terms of their existence and validity periods; these do not explicitly capture the changes in relationship types. Other models such as TERC+ [50] capture the temporal nature of relationship types by expressing the relationship changes in terms of entity transformations. This model basically uses entity subtypes to represent temporal evolution of entities as well as relationships and hence might not be able to represent evolving relationships between entities without subtypes.

In this chapter a spatio-temporal network model named time aggregated graph [16, 17] is described. Time-aggregated graph, models the changes in a spatio-temporal network by collecting the node/edge attributes into a set of time series. The model can also account for the changes in the topology of the network. The edges and nodes can disappear from the network during certain instants of time and new nodes and edges can be added. The time-aggregated graph keeps track of these changes through a time series attached to each node and edge that indicates their presence at various instants of time. The representational capability of the model is illustrated through various application domains such as transportation science and emergency planning. The model is compared with another graph-based model, the time expanded graph, in the context of various application domains.

2.1.1 Illustrative Application Domains

Transportation networks are the kernel framework of many advanced transportation systems such as the Advanced Traveler Information System and Intelligent Vehicle Highway Systems. Transportation networks are spatio-temporal in nature and require significant database support to handle the storage of their large amounts of multi-dimensional data. Many important applications based on transportation networks, including travelers' trip planning, consumer business logistics, and

Table 2.1 Example Queries with Time-variance and Flow Networks

	Static	Time-variant
Graph (no capacity constraints)	Which is the shortest travel time path from Minneapolis downtown to airport?	Which is the shortest travel time path from Minneapolis downtown to airport at different times of a work day?
Flow network	What is the capacity of Twin-Cities freeway network to evacuate Minneapolis downtown?	What is the capacity of Twin-Cities freeway network to evacuate Minneapolis downtown at different times in a work day?

evacuation planning need to be built upon spatio-temporal network databases. For example, commuters try to find a suitable time to start their commute so that they spend the least time in traffic. There are many factors affecting the start time and the shortest route such as congestion levels, incident location, and construction zone. This is illustrated by the simple time-variant network shown in Fig. 2.2. It can be seen that the travel time from node N1 to node N2 changes with the start time. If the travel starts at $t = 1$, the commute time would be 6 units; travel on the same route would take 4 units if the start time is moved to $t = 3$. This shows that the shortest paths in a time-dependent network vary with time which adds a new dimension to shortest path computation which cannot be ignored. With the increasing use of sensor networks to monitor traffic data on spatial networks and the subsequent availability of time-varying traffic data, it becomes important to incorporate this data into the models and algorithms related to transportation networks. One of the greatest challenges in transportation science is how to manage traffic in time-varying transportation networks, especially in disaster situations. Popular models of emergency traffic use time-variant flow-network [1] operations like min-cut and max-flow [5]. The queries typically encountered in emergency traffic management would involve time-variant properties, as illustrated in Table 2.1.

In crime analysis and prevention, identifying the areas of increasing criminal activity is a key step. Computing the routes that show significantly high crime rates can improve the efficiency of the patrol operations. Crime data usually consists of the geographical location of the crime, type of crime and its time of occurrence [28]. To compute the routes of high criminal activity, a model is required to represent the underlying transportation network along with the time dependent crime data associated with its edges and nodes. For example, the crime rates can vary with the time of the day and the interesting routes can change. With the availability of time-varying data, it becomes important to incorporate this data in the models and analysis of crime data.

Another interesting area of exploration is the effect of temporal dimension on conceptual models such as Entity-Relationship (ER) model [3] and more specifically

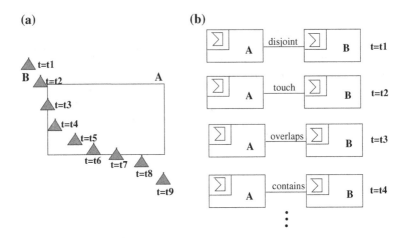

Fig. 2.1 Illustration of a dynamic relationship between two objects and its representation. **a** Location of Moving Object. **b** PEER Diagrams for (a)

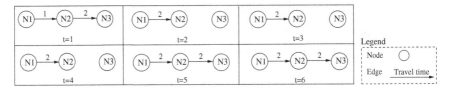

Fig. 2.2 Network at various instants

on the Pictogram-Enhanced Entity-Relationship (PEER) diagram [46]. A simple example is shown in Fig. 2.1. It illustrates a scenario where a moving sensor B crosses a geographic area A. Figure 2.1a shows the locations of B at discrete time instants ($t = t_1, t_2, t_3, t_4, t_5, t_6, t_7, t_8, t_9$). The relationship of object B with object A changes with time. This has been represented in Fig. 2.1b using a series of PEER diagrams. Each diagram represents the relationship at an instant. For example, the first diagram represents the time instant $t = t_1$ when the relationship between the objects is 'disjoint'. The figure shows the representations for the first four instants; the rest are modeled in a similar manner.

2.1.2 Broad Computer Science Challenges

A time-variant graph is a graph whose edge and node properties and topological structure are time dependent. For example, traffic volume on urban highways varies over the time of day which leads to variation in travel time. In addition to network parameter values, the network topology can also change with time due to the unavailability of certain road segments during some periods of time due to repair or natural calamities. There are also be cases where the road segments are unavailable periodically due to traffic management strategies such as using all lanes of a street in the same direction to handle peak time congestion. Conventional graph algorithms cannot easily be applied to the snapshots at discrete time instants to evaluate frequent queries without accounting for relationships among snapshots.

Time-variant graphs raise many challenges for database research. Due to their potentially large and evergrowing sizes, a storage-efficient representation is critical to reduce and possibly eliminate redundant information across different time-points. Second, new data model concepts need to be investigated to represent and classify potentially new alternative semantics for common graph operations such as shortest-path and connectivity. For example, a shortest path between a given pair of nodes may have at least two interpretations, one for a given start time-point and the other for the shortest travel-time for any start time in a given time interval. A third challenge is the design of efficient and correct query processing strategies and algorithms since some of the commonly assumed graph-properties may not hold for spatio-temporal graphs. For example, consider the optimal prefix property (a requirement for the greedy approaches [5]) for shortest paths in a graph. While each prefix path (path from a source node to an intermediate node in an optimal path) is optimal in a static graph, it may not be optimal in a spatio-temporal graph due to the potential wait at the intermediate node. In the network shown in Fig. 2.2, the best time to start a journey from node N1 to node N3 is $t = 4$, which takes 4 time units. The optimal path from N1 to N3 that starts at $t = 4$ is not optimal for the intermediate node N2. The best start time for a path from N1 to N2 is $t = 1$, which proves to be sub-optimal for a journey from N1 to N3. The lack of optimal prefix property in best start time shortest paths rules out the possibility of using a greedy strategy in algorithm design.

Key Features Of TAG:

Graph Aggregation: The temporal variation in the topology and parameter values can be represented using aggregates as edge/node attributes in the graph used to represent the spatial network. The edges and nodes can disappear from the network during certain instants of time and new nodes and edges can be added. The time-aggregated graph keeps track of these changes through a time series attached to each node and edge that indicates their presence at various instants of time.

Query Language: A query language needs to represent common queries. A key challenge is to define a complete set of logical operators for the time-aggregated graph.

Query Processing: The time aggregated graph with the proposed query operators will be used to process queries pertaining to the domain applications. A frequent query that arises in spatio-temporal networks is the shortest path computation. The algorithm needs to consider the availability of the required edges and nodes at the appropriate time instants. If the shortest route and the shortest route travel time are time-dependent, shortest path computation can be performed for a given start or it can find the least travel time path over the entire time period of interest.

In this chapter we describe a model for spatio-temporal networks called the time aggregated graph based on graph aggregation. The time-aggregated graph keeps track of the time-dependence of a graph through a time series attached to each node and edge that indicates their presence at various instants of time. We show that this model has less storage requirements than time expanded networks since it does not rely on replication of the entire network across time instants. We define a set of logical operators based on the time aggregated graph.

2.2 Basic Concepts

Spatial networks that show time-dependence serve as the underlying networks for most location based services. Traditionally graphs have been extensively used to model spatial networks (e.g. road networks) [40]; weights assigned to nodes and edges are used to encode additional information. In a real world scenario, it is not uncommon for these network parameters to be time-dependent. Formulation of computationally efficient and correct algorithms for the shortest path computation that takes into account the dynamic nature of the networks is important. Models of these networks need to capture the possible changes in topology and values of network parameters with time and provide the basis for the formulation of computationally efficient and correct algorithms for the frequent computations like shortest paths. Given the set of frequent queries posed by an application on a spatial network and the patterns of variations of the spatial network with time, we need to find a model that supports efficient and correct algorithms for computating the query results, while trying to minimize the storage and cost of computation. In this section we discuss the basics of the model used to represent spatial networks called "time aggregated networks" [16]. The algorithm presented in this paper is formulated based on this model. Time aggregated graphs can not only capture the time-dependence of network parameters, but also account for the possibility of edges and nodes being absent during certain instants of time.

2.2.1 The Conceptual Model

A graph $G = (N, E)$ consists of a finite set of nodes N and edges E between the nodes in N. If the pair of nodes that determine the edge is ordered, the graph is directed; if it is not, the graph is undirected. In most cases, additional information is attached to the nodes and the edges. In this section, we discuss how the time dependence of these edge/node parameters are handled in the proposed model, the time-aggregated graph.

We define the time-aggregated graph as follows.

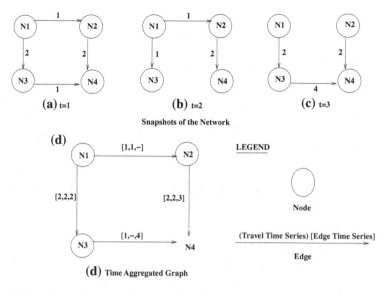

Fig. 2.3 Network at various time instants and the time aggregated graph

$taG = (N, E, TF, f_1 \ldots f_k, g_1 \ldots g_l, w_1 \ldots w_p | f_i : N \rightarrow \mathbb{R}^{TF}; g_i : E \rightarrow \mathbb{R}^{TF}; w_i : E \rightarrow \mathbb{R}^{TF})$ where

N is the set of nodes,

E is the set of edges,

TF is the length of the entire time interval,

$f_1 \ldots f_k$ are the mappings from nodes to the time-series associated with the nodes,

$g_1 \ldots g_l$ are mappings from edges to the time series associated with the edges, and

$w_1 \ldots w_p$ indicate the time dependent weights (eg. travel times) on the edges.

Each edge has an attribute, called an edge time series that represents the time instants for which the edge is present. This enables the time aggregated graph to model the topological changes of the network with time. It is assumed that each edge travel time has a positive minimum and the presence of an edge at time instant t is valid for the closed interval $[t, t + \sigma]$.

Figure 2.3a–c shows a network at three time instants. The network topology and parameters change over time. For example, edge N3–N4 is present at time instants $t = 1, 3$, and disappears at $t = 2$ and its weight changes from 1 at $t = 1$ to 4 at $t = 3$. The time aggregated graph that represents this dynamic network is shown in Fig. 2.3d. In this figure, edge N3–N4 has two attributes, both being a series. The attribute $(1, 3)$ represents the time instants at which the edge is present and $[1, \infty, 4]$ is the weight time series, indicating the weights at various instants of time. Figure 2.4a shows the time aggregated graph (corresponding to Fig. 2.3a–c and the time expanded graph that represent the same scenario. Edge weights in a time expanded graph are not explicitly shown as edge attributes; instead they are represented by edges that connect the copies of the nodes at various time instants. For example, the weight

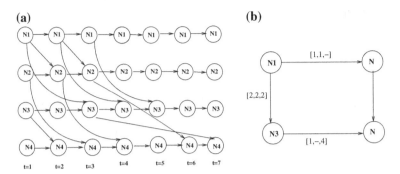

Fig. 2.4 Time-aggregated graph versus time expanded graph. **a** Time Expanded Graph. **b** Time-aggregated Graph

1 of edge N1–N2 at $t = 1$ is represented by connecting the copy of node N1 at $t = 1$ to the copy of node N2 at time $t = 2$. The time expansion for the example network needs to go through 7 steps since the latest time instant would end in the network is at $t = 7$. For example, the traversal of edge N3–N4 that starts at $t = 3$ ends at $t = 7$, the travel time of the edge being 4 units. The number of nodes is larger by a factor of T, where T is the number of time instants and the number of edges is also larger in number compared to the time-aggregated graph. If the value of T is very large in a spatial network, it would result in enormously large time expanded networks and consequently slow computations.

2.2.2 A Logical Data Model

Basic Graph Operations

We extend the logical data model described in [40] to incorporate the time dependence of the graph model. The framework of the model consists of two dimensions (1) graph elements, namely node, edge, route and graph and (2) operator categories that consist of accessors, modifiers and predicates. A representative set of operators for each operator category is provided in Tables 2.2, 2.3 and 2.4. Table 2.2 lists a representative set of 'access' operators. For example, the operator *getEdge(node1,node2,time)* returns the edge properties of the edge from node *node1* to node *node2*, such as the edge identifier (if any) and associated parameters at the specified time instant. For example operator *getEdge(N1,N2,1)* on the time-aggregated graph shown in Fig. 2.3 would return the travel time of the edge N1–N2 at $t = 1$, that is 1. Similarly, *get_edge(node1,node2)* returns the edge properties for the entire time interval. In Fig. 2.3, the operator *get_edge(N1,N2)* would result in $(1, 1, \infty)$. *get_edge_earliest(N3,N4,2)* returns the earliest time instant at which the edge N3–N4 is present after $t = 2$ (that is $t = 3$). Table 2.3 shows a set of modifier

Table 2.2 Examples of operators in the accessor category

	at_time	at_all_time	at_earliest
Node	get(node,time)	get_node(node)	get_node_earliest_Presence (node,time)
Edge	getEdge(node1,node2,time)	get_edge(node1,node2)	get_edge_earliest_Presence (node1,node2,time)
Route	getRoute(node1,node2,time)	get_route(node1,node2)	get_route_earliest_Presence (node1,node2,time)
Graph	get_Graph(time)	get_Graph()	–

Table 2.3 Examples of operators in the modifier category

	Insert		Delete		Modify	
	at_time	at_all_time	at_time	at_all_time	at_time	at_all_time
Node	insert(node, time,value)	insert(node, valueseries)	delete(node, time)	delete(node) delete(node)	update(node, time,value)	update(node, valueseries)
Edge	insert(node1, node2, time,value)	insert(node1, node2, valueseries)	delete(node1, node2, ,time)	delete(node1, ,node2) ,node2)	update(node1, node2,time value)	update(edge, valueseries)
Route	insert(node1 node2,time)	insert(node1 ,node2)	delete(node1 ,node2,time)	delete(node1, node2)	– –	
Graph	insert(graph time)	insert(graph) insert(graph)	delete(graph, time)	delete(graph) delete(graph)	update(graph, ,time)	update(graph)

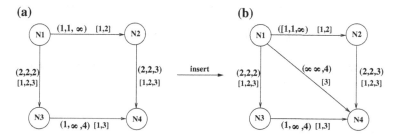

Fig. 2.5 A time aggregated graph before and after an insert operation. **a** Before insert, **b** after insert

operators that can be applied to the time aggregated graphs. For example, Fig. 2.5a, b show a time aggregated graph before and after the *insert(N1,N4,3,4)* operation this operation inserts edge N1–N4 at time instant $t = 3$ and the edge cost is 4. We also define two predicates on the time-aggregated graph.

exists_at_time_t: This predicate checks whether the entity exists at the start time instant t.
exists_after_time_t: This predicate checks whether the entity exists at a time instant after t.

Table 2.4 Predicate operators in time-aggregated graphs

	exists_at_time_t	exists_after_time_t
Node	exists(node u,at_time_t)	exists(node u,after_time_t)
Edge	exists(node u,node v, at_time_t)	exists(node u,node v, after_time_t)
Route	exists(node u,node v,a_route r at_time_t)	exists(node u,node v,a_route r, after_time_t)

Table 2.4 illustrates these operators. For example, node v is adjacent to node u at any time t if and only if the edge (u, v) exists at time t as shown in the table. *exists(N1,N2,1)* on the time aggregated graph in Fig. 2.3 returns a "true" since the edge N1–N2 exists at $t = 1$.

The fundamental entities in graphs, namely, *Graph*, *Node*, and *Edge* and a series of common operations that are associated with each class are listed.

```
public class Graph    {
   public void add(Object label, timestamp t);
   // node with the given label is added at the time
   instant t.

   public void addEdge(Object n1, Object n2, Object
   label, timestamp t, timestamp t_time)
   // an edge is added with start node n1 and end node
      n2 at
   // time instant t and travel time, t_time.

   public Object delete(Object label, timestamp t)
   // removes a node at time t and returns its label.

   public Object deleteEdge(Object n1, Object n2,
   timestamp t)
   // deletes the edge from node n1 to node n2 at t.

   public Object get(Object label, timestamp t)
   // returns the label of the node if it exists at
      time t.

 public Iterator get_node_Presence_Series(Object n1)
 // the   presence series of  node n1 is returned.

 public Object getEdge(Object n1, Object n2, timestamp
 t)
 // returns the edge from node n1 to node 2 at time
```

instant t.

```
public Iterator get_edge_Presence_Series(Object n1,
Object n2)
// the  presence series of edge from node n1 to node
   n2
// is returned.
```

```
public Object get_a_Successor_node(Object label,
timestamp t)
//  an adjacent node of the vertex is returned if an
    edge exists
//  to this node at a time instant at or after t.
```

```
public Iterator get_all_Successor_nodes(Object label,
timestamp t)
//  all adjacent nodes are returned if edges exist to
    them
//  at time instants at or after t.
```

```
public Object get_an_earliest_Successor_node(Object
label,timestamp t)
//  the adjacent node which is connected to the given
    node with
//  the earliest time stamp after t is returned.
```

```
public timestamp get_node_earliest_Presence(Object
n1, timestamp t)
//  the earliest time stamp after t at which the node
    n1
//  is available is returned.
```

```
public timestamp get_node_Presence_after_t(Object n1,
timestamp t)
//  Part of the presence time series of node n1 after
    time t
//  is returned.
```

```
public timestamp get_edge_earliest_Presence(Object
n1, Object n2, timestamp t)
//  the earliest time stamp after t at which the edge
    from
// node n1 to node n2 is available is returned.
```

```
public timestamp get_edge_Presence_after_t(Object n1,
Object n2, timestamp t)
//  Part of the presence time series of edge(n1-n2)
    after time t
//  is returned.


}
```

A few important operations associated with the classes **Nodes** and **Edges** are p
rovided below.

```
public class Node  {
        public Node(Object label, timestamp t)
        // the constructor for the class. A node with
           the appropriate
        // label is created at the time t.

        public Object label()
        // returns the label associated with the node
           if it exists at t.
}

public class Edge  {
        public Edge(Object n1, Object n2, Object
        label, timestamp t_inst, timestamp t)
        // the constructor for the class. an edge is
           added with start
        // node n1 and end node n2 at time instant t
           and
        // travel time, t_time.

        public Object start()
        // returns the start node of the edge.

        public Object end()
        // returns the end node of the edge.
}
```

2.2.3 Physical Data Model

A static graph $G = (V, E)$ can be represented using an adjacency matrix. This is a
$|V| \times |V|$ matrix, A such that the element a_{ij} is defined as
$a_{ij} = w_{ij}$ if $ij \in E$ and w_{ij} is the weight of the edge ij and

$a_{ij} = 0$, otherwise. This representation requires $O(N^2)$ memory. It can be seen that the storage required for this representation is independent of the number of edges in the graph, in relation to the number of nodes. In other words, there is no saving in memory even when the graphs are sparse. One representation that can exploit such sparsity is the adjacency list representation. The adjacency list representation of a graph $G = (V, E)$ consists of an array of lists, one for each vertex $v \in V$. The list corresponding to a vertex v contains all vertices that are adjacent to v in G. For a directed graph, the space requirement for the lists is $O(m)$ where $m = |E|$. The total memory requirement is $O(n + m)$ where $n = |V|$. The weight of each edge uv is stored with the vertex v in u's adjacency list. This representation is especially suitable for sparse graphs.

Time aggregated graphs can be represented by either one of the representation, with the necessary modifications. These representations need to be extended to include the time series representations on edges (corresponding to time dependent edge costs) and nodes. Adjacency list representation is extended by adding a list to each vertex in the adjacency list. Adjacency list representation uses an array of pointers one pointer for each node. The pointer for each node points to a list of immediate neighbors. Stored at each neighbor node are the edge presence series and travel times for the edge starting from the first node to this neighbor. Since the length of the time series is T, where T is the length of the time period, the adjacency list representation would require $O(m + n + nT + mT)$, where n is the number of nodes and m is the number of edges. In reality, not all time series would be of length T and assuming an average length α, the storage would be $O(n + m + \alpha n + \alpha m)$. The time series store a single value if the value of the attribute remains constant, indicated by the character 'F'. If the value of the attribute changes over time, it is indicated by the character 'V'.

To extend the adjacency matrix to represent the time aggregated graph, a third dimension can be added. The new matrix A would be $n \times n \times T$, requiring $O(n^2 T)$ memory. Figure 2.6a, b show the adjacency list and adjacency matrix representations for the time aggregated graph shown in Fig. 2.3. For example, the edge N1–N2 in the graph at $t = 1$ is represented by the pointer from N1 to N2 in the adjacency list. The array $(1, 2, \infty)$ is stored at N2 to represent the travel times at $t = 1, 2, 3$ for the edge N1N2. In the adjacency matrix the presence of edge N1N2 at a time instant $t = 1$ is represented by $A[1, 2, 1] = 1$, since the travel time for the edge is 1 unit at $t = 1$. Since the edge is absent at an instant $t = 3$, $A[1, 2, 3] = \infty$ which indicates an infinite edge cost at time instant $t = 3$. Note that the start node, the end node and the time instant are represented by the first, second and the third dimension of the matrix. Though the adjacency matrix has been illustrated as three snapshots in Fig. 2.6b for the sake of clarity, they are represented in one, three-dimensional matrix.

Logical operations on a time-aggregated graph can be classified as

1. Topology first operators (graph dominated operations).
   Examples include get_route(n1,n2) and get_edge(n1,n2).
2. Time-first operators (Time dominated queries).

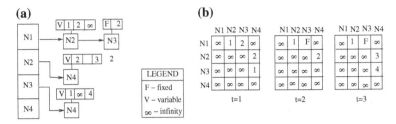

Fig. 2.6 Storage structures for time aggregated graph. **a** Adjacency list representation, **b** adjacency matrix representation

Some examples are get_Graph(time t) and get_edge_at_t(n1,n2,t).

Both representations are equally capable of handling graph dominated queries. To compute time first operations (snapshot queries such as to find the graph at a given time instant), adjacency matrix representation is more suitable. In this representation, these queries represent the time slices of the matrix at the given time instants.

Graphs representing transportation networks are generally sparse and hence adjacency list representation is more likely to be storage efficient compared to adjacency matrix representations. The choice is hence a tradeoff between the storage cost and the frequency of time dominated queries. We expect route queries (which are topology first queries) to be more frequent and since adjacency list representation is capable of handling these, based on storage costs, we used adjacency lists in our implementations. Moreover, most databases use adjacency list representation.

2.2.3.1 Towards Handling Infinite Time Series

In most domains that involve spatio-temporal networks such as transportation networks, crime data analysis, and sensor networks data is continuously collected at discrete instants of time. For example, sensors on urban highways measure congestion levels every 30 s and crime data is appended with every time a crime occurs.. Conceptually, the time aggregated graph can be viewed as a time series of graphs. Each graph represents the attribute values and the topological structure of the network at the given instant of time. Based on the periodicity of data collection, the application domains can be broadly classified into (1) applications where data is measured periodically and (2) applications such as crime analysis where data is recorded when an event occurs.

When data is measured periodically, the underlying model should be able to capture the changes that take place in the spatio-temporal network at every instant. Time aggregated graphs represent this as a time series of graphs, each graph in the series modeling the state of the network. For example, the state of a road network at $t = t_1$ would be represented as a graph corresponding to this instant. The state of a sensor network, which would include the measurements at an instant would also be modeled in a similar manner.

(a)   **(b)**

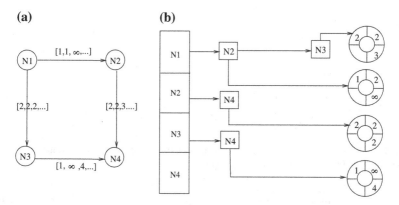

Fig. 2.7 Representation of sliding windows in time aggregated graph. **a** Time aggregated graph, **b** implementation of time series

In application domains where the network state changes due to an event, the time aggregated graph stores the tuples of time stamp and the event.

Implementation

The time series of graphs would be implemented as a graph where the node and edge attributes are time series. Most application domains deal with 'infinite' streams of data, and the edge and node attributes are possibly infinite time series. One implementation uses sliding windows implemented through circular buffers. Figure 2.7a shows a time aggregated graph with time series attributes on its edges. Figure 2.7b shows the modified adjacency list representation that implements an infinite time series. Each time series is stored in a circular buffer.

2.3 Evaluation and Validation

2.3.1 Representational Comparison: Time Aggregated Graphs Versus Existing Models

A time-expanded network has one copy of the set of nodes for each discrete time instant. Corresponding to each edge with transit time t in the original network, there is a copy of an edge (called the cross edge) between each pair of copies of nodes separated by the transit time t [14, 22, 26]. Thus, a time-dependent flow in a dynamic network can be interpreted as a static flow in a time expanded network. This allows application of static algorithms on such networks to solve dynamic flow problems. Apart from the "enormous increase in the size of the underlying network" [26] the suitability of the model in some application domains needs further exploration.

A time expanded graph assumes that the edge weight represents a flow parameter, and it represents the time taken by the flow to travel from the source node to the end node. This is represented by the cross edges between the copies of the graph.

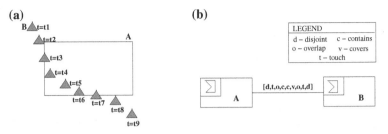

Fig. 2.8 Illustration of a dynamic relationship between two objects and its representation. **a** Locations of Moving Object. **b** TAG Representation for (a)

Since the cross edges in a time expanded graph represent a flow across the nodes, the representation of non-flow networks using this model is not obvious. By contrast, the time aggregated graph model does not impose such a restriction because the attributes are collected into a time series. This difference can be illustrated through the example of the possible extension of the PEER diagram explained in Sect. 2.1.1. While time aggregated graph would model the time-dependent relationships as a time series on the edge connecting the nodes (that represent the entities), the representation of the same scenario is not obvious when time expanded graphs are used. An illustration of the representation of time-dependent relationships using time-aggregated graph representation for the scenario depicted in Fig. 2.1 is shown in Fig. 2.8. Figure 2.1 shows the locations of B at discrete time instants ($t = t_1, t_2, t_3, t_4, t_5, t_6, t_7, t_8, t_9$). The relationship of object B with object A changes with time. This has been represented in Fig. 2.8a using an aggregated representation. The line segment that represents the relationship has an attribute which is an ordered set, each element indicating the current relationship of object B with A. For example, the second entry 'o' indicates that the object B touches A at $t = t_2$ and overlaps A at $t = t_3$. In the domain of crime analysis, the number of crimes reported on a road segment (represented by an edge) at a given time might not be meaningfully represented by an edge in the time expanded graph. The time aggregated graph would represent this as an element in its time series attribute.

In most spatio-temporal networks, the length of the time period (indicated by T in this paper) might not be known in advance since data arrives as a sequence at discrete time instants. For example, sensors in transportation networks collect data at a rate of about once every 30 s. Crimes are reported whenever an incident occurs. In addition to being able to represent these attributes, the model must be capable of handling infinite sequences of data. Since time expanded networks require a prior estimate of the length of the time period T, handling of infinite time series might not be easy and obvious. Also, the necessity for the prior knowledge of T might lead to problems in the algorithms based on time expanded networks since an underestimation of T can result in failure of finding a solution. On the other hand, an over-estimated T will result in an over-expanded network and hence lead to unnecessary storage and run-time and would adversely affect the scalability of the algorithms.

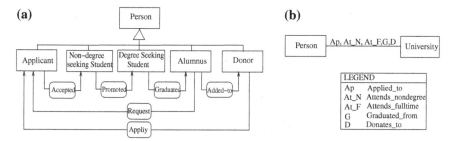

Fig. 2.9 Representations of dynamic relationships in TERC and aggregated graph. **a** An example
TERC model, **b** aggregated model (**a** adapted from [50])

Time expanded graphs model the time-dependence of edge parameters through
the cross edges that connect the copies of the nodes. This representation, thus, does
not provide the means to separate data (for example, an edge attribute series) from its
physical representation and hence can adversely affect physical data independence.

The temporal conceptual model TERC+ [50] models dynamic relationships
between entities using evolutions of the entities involved. The temporal nature is cap-
tured through representing transitions of objects. An example is shown in Fig. 2.9. It
represents a dynamic relationship between a person and a University. The relationship
changes from an applicant to a donor after graduation. The change in the relationship
is represented through various classes of the same entity as shown in Fig. 2.9a. An
aggregated model of the same scenario is shown in Fig. 2.9b. Though at the finest
level, the representations would be the same, the aggregated model facilitates a better high level summarization. This model might not be sufficient to represent cases
where entity subtypes cannot be used to model evolving relationships. For example,
Fig. 2.1, represents a scenario where the entities (a sensor and a geographic area)
involved in the dynamic relationship do not have subtypes and hence might not yield
itself to this model.

2.3.2 Comparison of Storage Costs with Time Expanded Networks

According to the analysis in [41], the memory requirement for a time expanded
network is $O(nT) + O(n + mT)$, where n is the number of nodes, m is the number
of edges in the original graph, and T is the length of the travel time series. The
framework of a time aggregated graph would require a memory of $O(n + m)$, where
n is the number of nodes and m is the number of edges. Each edge that has a time-
varying attribute has an attribute time series associated with it. If the average length
of the time series is $\alpha(\leq T)$, the memory required is $O(\alpha m)$, assuming an adjacency
list representation. The total memory requirement for a time aggregated graph is
$O(n + m + \alpha m)$. This comparison shows that the memory usage of time-aggregated
graphs is less than that of time expanded graphs $nT > n$ and $\alpha \leq T$.

Fig. 2.10 TAG: storage costs

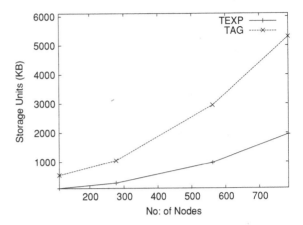

Table 2.5 Description of datasets

Dataset	Radius (miles)	Number of nodes	Number of edges
1	0.5	111	287
2	1	277	674
3	2	562	1443
4	3	786	2106

Experimental Evaluation:

Figure 2.10 shows the result of evaluation of storage requirement for time aggregated graph in comparison with time expanded graph. The networks used are the road maps from the Minneapolis downtown area, of radii, 0.5, 2 and 3 miles. This is appended with travel time series of length 200. The number of nodes and edges in these datasets are provided in Table 2.5.

The evaluation shows that memory cost of TAG is much less than the time expanded graph.

2.4 Summary

Spatio-temporal networks form a key part of critical applications such as emergency planning and there is a great need for database support in this area. This chapter describes a model based on time aggregation to represent a spatio-temporal network. Time aggregated graphs represent the time variant properties by aggregating edge and node parameters into time series. The analytical and experimental analysis of storage cost requirements of time aggregated graph are presented.

Chapter 3
Shortest Path Algorithms for a Fixed Start Time

Abstract Shortest path computation is an important query on any network. In a spatio-temporal network, this computation assumes added semantics due to the dependence of network attributes on time. Shortest paths can be computed either for a given start time or to find the start time and the path that leads to least travel time journeys (best start time journeys). Developing efficient algorithms for computing shortest paths in a time varying spatial network is challenging because these journeys do not always display greedy property or optimal substructure. This chapter describes algorithms to compute shortest paths for a given start time. The formulations of shortest path algorithms can also depend on the properties of the network parameters such as travel times. For example, the algorithm can significantly vary depending on whether the travel times follow FIFO property or not. The chapter provides algorithms for both FIFO and non-FIFO travel times.

3.1 Introduction

Spatio-temporal networks are spatial networks whose topology and parameters change with time. These networks are important due to many critical applications such as emergency traffic planning and route finding services and there is an immediate need for models that support the design of efficient algorithms for computing the frequent queries on such networks. This problem is challenging due to potentially conflicting requirements of model simplicity and support for efficient algorithms. Time expanded networks which have been used to model dynamic networks employ replication of the network across time instants, resulting in high storage overhead and algorithms that are computationally expensive. In contrast, time-aggregated graphs do not replicate nodes and edges across time; rather they allow the properties of edges and nodes to be modeled as a time series. Since the model does not replicate the entire graph for every instant of time, it uses less memory and the algorithms for common operations (e.g. connectivity, shortest path) are computationally more

B. George and S. Kim, *Spatio-temporal Networks*,
SpringerBriefs in Computer Science, DOI: 10.1007/978-1-4614-4918-8_3,
© The Author(s) 2013

efficient than those for time expanded networks. One important query on spatio-temporal networks is the computation of shortest paths. Shortest paths can be computed either for a given start time or to find the start time and the path that leads to least travel time journeys (best start time journeys). Developing efficient algorithms for computing shortest paths in a time varying spatial network is challenging because these journeys do not always display greedy property or optimal substructure, making techniques like dynamic programming inapplicable. The formulations of shortest path algorithms can also display different characteristics based on some properties of the network parameters such as travel time.

3.1.1 Broad Challenges

A time-variant graph is a graph whose edge and node properties and topological structure are time dependent. For example, traffic volume on urban highways varies over the time of day, which leads to a variation in travel time. In addition to network parameter values, the network topology can also change with time due to the unavailability of certain road segments during some periods of time due to repair or natural calamities. Conventional graph algorithms cannot easily be applied to the snapshot graphs at discrete time instants to evaluate frequent queries without accounting for relationships among snapshots. However, time-variant graphs raise many challenges for database research. Due to their potentially large and evergrowing sizes, a storage-efficient representation is critical to reduce and possibly eliminate redundant information across different time-points. Second, new data model concepts need to be investigated to represent and classify potentially new alternative semantics for common graph operations such as shortest-path and connectivity. For example, a shortest path between a given pair of nodes may have at least two interpretations, one for a given start time-point and the other for the shortest travel-time for any start time in a given time interval. A third challenge is the design of efficient and correct query processing strategies and algorithms since some of the commonly assumed graph-properties may not hold for spatio-temporal graphs. For example, consider the optimal substructure (required in dynamic programming, [5]) for shortest paths in a graph. While each prefix path (path from a source node to an intermediate node in an optimal path) is optimal in a static graph, it may not be optimal in a spatio-temporal graph due to a potential wait at an intermediate node.

This chapter presents a classification of shortest path algorithms based on the start time choice and the characteristics of travel time variations (FIFO or non-FIFO). It further proceeds to describe two shortest path algorithms for a given start time in a network where travel times follow FIFO property. An algorithm that computes the shortest path for a given start time in a non-FIFO network is discussed. The non-FIFO shortest path algorithm is based on a transformation called arrival time series transformation (ATST) that converts a non-FIFO to a network that displays optimal prefix property leading to the formulation of a greedy algorithm for a non-FIFO network.

3.2 Basic Concepts

This sections presents a classification of shortest path algorithms and lists some of the algorithmic challenges.

3.2.1 Classification of Shortest Path Algorithms

One of the most frequent queries on any spatio-temporal network is the computation of shortest paths. In time dependent networks, the shortest path and the traversal time are dependent on the start time. For example, a shortest path from node N1 to node N5 in Fig. 3.1 for the start time $t = 1$ takes a travel time of 6 units. If the start is postponed to $t = 3$, the travel time drops to 4 units. Due to the time dependence of shortest paths, in a spatio-temporal network it is possible to raise interesting queries such as "When is the best time to start a journey so that time spent in the network is the least?". Sections 3.1 and 3.4 describe the formulations of two shortest path problems, first for a fixed start time and second, for the least travel time. The design of these algorithms effectively utilizes certain properties of the time dependent parameters (such as the FIFO property of travel time). The classification of the algorithms is shown in Fig. 3.2. The shortest path algorithms show significantly different properties based on their formulations. For example, shortest path computation for a fixed start time might display optimal prefix property under certain assumption on travel time characteristics. Computation of best start time shortest path under non-FIFO travel times might not always display optimal prefix property ruling out popular design techniques such as greedy and A* based algorithms. Based on these characteristics, shortest path computation on time aggregated graph falls under various categories as shown in Fig. 3.2.

3.2.2 Algorithmic Challenges

A time dependent graph might not display some properties that would make some common algorithm design techniques such as dynamic programming and greedy strategy feasible. For example most time dependent graphs do not exhibit optimal prefix property, thus making it impossible to apply greedy methods in shortest path computations. Figure 3.3 shows a time dependent network represented as a time aggregated graph. The following example illustrates a scenario where Dijkstra's algorithm which uses greedy strategy to compute the shortest path identifies a non-optimal route. If the edge costs are assumed to be edge travel times, the cost of a node indicates the arrival time at the node. When the least cost node is expanded, the costs of the outgoing edges are chosen as the costs at the arrival time. In cases where an edge is not available, the cost at the earliest available time is selected.

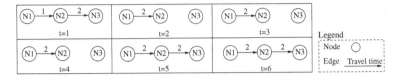

Fig. 3.1 Snapshots of a network

Fig. 3.2 Classification of TAG-based shortest path algorithms

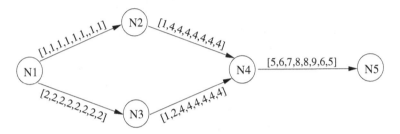

Fig. 3.3 Illustration of shortest path computation

Table 3.1 Trace of Dijkstra's algorithm for the network shown in Fig. 3.3

Iteration	N1	N2	N3	N4	N5
1	1 (closed)	∞	∞	∞	∞
2	1	2 (closed)	3	∞	∞
3	1	2	3 (closed)	6	∞
4	1	2	3	6 (closed)	15
5	1	2	3	6	15 (closed)

Table 3.1 shows the execution trace of Dijkstra's algorithm on the time aggregated graph shown in Fig. 3.3. The schedule in Table 3.2 shows the start time and arrival at each node as computed by the algorithm. As Table 3.2 illustrates, the algorithm computes the route N1–N2–N4–N5 as the shortest path for the start time $t = 1$ at node N1. The total travel time for the route is 14 time units. It can be seen that there is another route N1–N3–N4–N5 for the same start time which takes a travel time of 12 time units. This is an optimal route from N1 to N5, which suggests that the algorithm might not always compute the shortest path.

Table 3.2 Start and arrival times at nodes

Route computed by algorithm						Optimal route				
Node	N1	N2	N3	N4	N5	Node	N1	N2	N3	N4	N5
Arrival time	1	2	–	6	15	Arrival time	1	–	3	7	13
Start time	1	2	–	6	15	Start time	1	–	3	7	13
Total travel time = 14						Total travel time = 12					

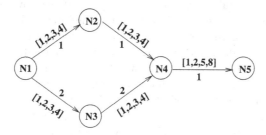

Fig. 3.4 Optimal sub-structure of shortest paths

3.3 Shortest Path Computation for Fixed Start Time

In time dependent networks, the shortest path and its traversal time are dependent on the start time at the source node. This section outlines the challenges encountered and the approaches to address them. It also provides an outline of the algorithm that computes the shortest path for a given start time in a time-dependent network. The algorithm uses the time aggregated graph to represent the network.

Challenges

(1) **Not all shortest paths for a given start time show optimal prefix property**:
The application of a greedy strategy in the shortest path computation (which is a popular choice in most optimization problems) in a time-aggregated graph faces a challenge. Not all shortest paths display the optimal sub-structure, as illustrated by Fig. 3.4. For the sake of simplicity, the travel times are constant in this example. It can be seen that a shortest path (N1–N3–N4–N5) from N1 to N5 for the start time $t = 1$, which takes 5 time units, does not display optimal prefix property. The path from N1 to N4 following the above path is not optimal (shortest path being N1–N2–N4). Although such paths that do not display optimal sub-structure could exist, it can be proved that there is at least one optimal path which satisfies the optimal sub-structure property.

Lemma 1 *If there is an optimal route from source to destination, then there is at least one optimal route from source to destination that shows optimal sub-structure.*

Proof As Fig. 3.4 illustrates, the failure of optimal structure of the shortest path occurs due to a potential wait at the intermediate node ($N4$), after reaching this node traversing the optimal path from $N1$ to $N5$, assuming travel time variations follow

FIFO property. Consider the optimal path from $N1$ to $N4$. Append this path to the path $N4$–$N5$ (allowing wait at the intermediate node $N4$) from the optimal path. This would be still the shortest path from $N1$ to $N5$. Otherwise, it would contradict the optimality of the original shortest path.

This result enables us to use a greedy approach to compute the shortest path.

(2) **Greedy approach in selecting the traversal time of an edge might not ensure correctness (optimality) of the shortest path**: If the travel times follow a "random" variation, a greedy choice on the edge traversal time (ie.,) selecting the edge at the earliest available time instant might not guarantee optimality. For example, consider computing the shortest path from node $N1$ to node $N3$ in Fig. 3.5 for a start time $t = 1$. Traversing edge $N2$–$N3$ as soon as $N2$ is reached (at $t = 2$) would result in a sub-optimal solution. Waiting at node $N2$ for a time unit and starting from node $N2$ at $t = 3$ would result in a total travel time of 3 units in comparison with 4 units if edge $N2$–$N3$ was traversed at $t = 2$. Shortest path algorithms for both FIFO and non-FIFO networks are described in the following sections. Algorithm that requires FIFO travel time (greedy version and A* search based) will be discussed in Sects. 3.3.1, 3.3.2 and the algorithm that can handle non-FIFO travel times will be discussed in Sect. 3.4.

(3) **Termination of algorithm is not guaranteed if there is a non-negative cycle over time:** If the graph has an infinite positive cycle and the travel times do not display FIFO property, an optimal path finding algorithm might not terminate since it will continue searching for a shortest path indefinitely.

Algorithms presented here assume finite time windows.

3.3.1 Shortest Path Algorithm for Fixed Start Time in a FIFO Network (SP-TAG)

The algorithm, called the SP-TAG algorithm, uses greedy strategy to find the shortest path for a fixed start time. Every node has a cost associated with it which represents the travel time to reach the node from the source node. The algorithm picks the node with the least cost and updates the costs of its adjacent nodes. While finding the adjacent nodes, each edge is selected at its earliest available time instant (min_t operation in the algorithm description). A trace of the algorithm is given in Table 3.3. The table entries are the costs associated with each node (representing the arrival times at the node) at each iteration. The node marked as "closed" is the node with the minimum cost selected for expansion. The travel times are assumed to follow the FIFO property.

Lemma 2 *The SP-TAG algorithm is correct.*

Proof The proof of correctness of the algorithm which follows a greedy strategy follows the proof of correctness for Dijkstra's algorithm to find the shortest path

---

Algorithm 1 Shortest Path (SP-TAG) Algorithm

Input:
1) $G(N, E)$: a graph G with a set of nodes N and a set of edges E;
     Each node $n \in N$ has a property:
          Node Presence Time Series : series of positive
integers;
          Each edge $e \in E$ has two properties:
               Edge Presence Time Series,
               Travel_time series : series of positive integers;
          $\sigma_{u,v}(t)$ - travel time of edge uv at time t.
2) s: Source node, $s \subseteq N$; 3) d: Destination node, $d \subseteq N$;
4) t_{start}: Start Time;
Output: Shortest Route from s to d for t_{start}
Method:
> $c[s] = t_{start}; \forall v \neq s, c[v] = \infty;$
> // $c[u]$ is the cost at the node u.
> Insert s in priority queue Q.
> while Q is not empty do {
> $u = extract\_min(Q);$
>   for each node v adjacent to u do {
> $t = min\_t((u, v), c[u]);$
>   if $t + \sigma_{u,v}(t) < c[v]$ {
> $c[v] = t + \sigma_{u,v}(t); \quad parent[v] = u;$
>   if v is not in Q, insert v in Q;
>   }
>   update Q;
>   }
> }
> }
> Output the route from s to d.

Table 3.3 Trace of the SP-TAG algorithm for the network shown in Fig. 3.4

Iteration	N1	N2	N3	N4	N5
1	1 (closed)	∞	∞	∞	∞
2	1	3 (closed)	3	∞	∞
3	1	3	3 (closed)	3	∞
4	1	3	3	4 (closed)	6
5	1	3	3	4	7 (closed)

Fig. 3.5 Illustration of non-FIFO paths

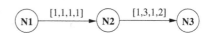

from a source node to a destination. The key difference in time aggregated graph is that each edge has a presence series. SP-TAG employs a greedy approach where it selects the earliest available time instant as the traversal time of the edge. Since waits are permitted at intermediate nodes, this admissible approach does not violate

Table 3.4 Trace of the SP-TAG_* algorithm for the network shown in Fig. 3.4

N1			N2			N3			N4			N5		
g()	h()	f()	g()	h()	f()	g()	h()	f()	g()	h()	f()	g()	h()	f()
1	4	5 (X)	∞	∞	∞	∞	∞	∞	∞	∞	∞	∞	∞	∞
1	4	5	3	3	6 (X)	3	4	7	∞	∞	∞	∞	∞	∞
1	4	5	3	3	6	3	4	7	4	2	6 (X)	∞	∞	∞
1	4	5	3	3	6	3	4	7	4	2	6	6	2	8

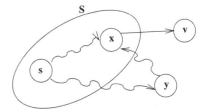

Fig. 3.6 Correctness of SP-TAG algorithm

the optimality of the shortest path even while considering the time-dependence of edge presence. To prove the correctness of the algorithm, it needs to be shown that the cost of a node, when it is closed, is the shortest path distance to the node. This can be proved by induction on the set of closed nodes (S in Fig. 3.6). Let v be the next node to be closed. Suppose the cost of node v was last updated when node x was added to S and v is adjacent to x. When x was added to S, a shorter path to v through x was discovered. Assume that the cost of v is not the shortest path cost. This would be due to the existence of a path $s \cdots y \cdots xv$ as shown in Fig. 3.6. Since x was closed before y, the shortest path to x is inside S by inductive hypothesis. Therefore, the length of the path from s to v through y cannot be shorter that the path $s \cdots xv$. The cost of v cannot be further reduced by forming a path through nodes outside S. Hence, the cost of the node when it is closed is the shortest path distance to the node.

Lemma 3 *The time complexity of the SP-TAG algorithm is $O(m(\log T + \log n))$ where T is the number of time instants, n is the number of nodes and m is the number of edges in the time aggregated graph.*

Proof The cost model analysis assumes an adjacency list representation of the graph with two significant modifications. The edge time series is stored in the sorted order. Attached to every adjacent node in the linked list are the edge time series and the travel time series.

For every node extracted from the priority queue Q, there is one edge time series look up and a priority queue update for each of its adjacent nodes. The time complexity of this step is $O(\log T + \log n)$. The asymptotic complexity of the algorithm would be

$$O(\Sigma_{v \in N}[degree(v).(\log T + \log n]) = O(m(\log T + \log n)).$$

The time complexity of the SP-TAG shortest path algorithm based on a time expanded network is $O(nT \log T + mT)$ [8]. Assuming a sparse graph where m is $O(n)$, $nT \log T < m \log T$. The SP-TAG algorithm is faster than the algorithm based on time expanded graph if $m \log n < mT$. In other words, the SP-TAG algorithm is faster if $\log n < T$.

3.3.2   A* Formulation of Shortest Path Algorithm for a Fixed Start Time in a FIFO Network (SP-TAG*)

The A* formulation of the shortest path algorithm for a fixed start time discussed in this section finds an optimal solution for FIFO travel times. The main challenge in formulating an A* search is the design of an admissible and monotone heuristic. We present the heuristic function used in the formulation and we prove the properties of the heuristic function which guarantee the optimality of the solution (admissibility) and the optimality of search (monotonicity).

Proposed Heuristic Function The evaluation function $f(n)$ of a node n is formulated as $f(n) = g(n) + h(n)$.

$g(n)$ is the actual cost to reach the node n from the start node s, which is the time taken to reach the current node from the start node. $h(n)$ is the estimated cost from the node n to a destination node d. We propose $h(n)$ to be the shortest path travel time from node n to the destination node d computed based the least travel time on each edge.

$$h(n) = Min_{t=1,2,\cdots,T} d_{ij}(t), \forall ij \in E$$

Lemma 4 *The heuristic function $h(n)$ is admissible.*

Proof A heuristic function $h(n)$ is admissible if it underestimates the cost from the node n to the destination node. Here, $h(n)$ is the shortest path from n to d based on the minimum travel times on each edge. Let S_{TAG} be the graph derived from a time aggregated graph where each edge cost time series has been replaced by a scalar cost equal to the minimum edge cost. Let P be the shortest path from node i to d in S_{TAG}.

Shortest path travel time $SP_{min} = \sum_{pq \in P} d_{pq}^{min}$

Let $P^*(t)$ be the shortest path in TAG that starts at node i at time t. For $h(n)$ to be admissible, $SP_{min} \leq SP(t)$.

$SP_{min} = \sum_{pq \in P} d_{pq}^{min} \leq \sum_{kl \in P^*} d_{kl}^{min}$ and $d_{kl}^{min} \leq d_{kl}(t)$.

$SP_{min} \leq \sum_{kl \in P^*} d_{kl}^{min} \leq \sum_{kl \in P^*} d_{kl}(t) = SP(t)$. The heuristic function is admissible.

Lemma 5 *The heuristic function $h(n)$ is monotone.*

Proof A heuristic function $h(n)$ is monotone if $h(i) \leq d_{ij} + h(j) \forall ij \in E$. Here, $SP_{id}^{min} \leq d_{ij}(t) + SP_{jd}^{min}$. $SP_{id}^{min} \leq d_{ij}^{min} + SP_{jd}^{min}$; else, it is a contradiction to the optimality of SP_{id}^{min}. Since $d_{ij}^{min} \leq d_{ij}^{(}t)$, $SP_{id}^{min} \leq d_{ij}(t) + SP_{jd}^{min}$.

Since the heuristic function is admissible and monotone, the A^* algorithm finds an optimal solution and performs and optimal search [24, 37]. A trace of the algorithm is given in Table 3.4. The table entries are the costs associated with each node (representing the arrival times at the node) at each iteration. The node marked as "X" is the node with the minimum cost selected for expansion. The travel times are assumed to follow the FIFO property.

Algorithm 2 A* based Shortest Path (SP-TAG_*) Algorithm

Input:
```
1) G(N, E): a graph G with a set of nodes N and a set of edges E;
       Each node n ∈ N has a property:
            Node Presence Time Series : series of positive
integers;
       Each edge e ∈ E has two properties:
            Edge Presence Time Series,
            Travel_time series : series of positive integers;
```
$\sigma_{u,v}(t)$ - travel time of edge uv at time t.
```
2) s: Source node, s ⊆ N; 3) d: Destination node, d ⊆ N;
```
4) t_{start}: Start Time;

Output: Shortest Route from s to d for t_{start}

Method:
```
       Preprocess: find the shortest path from node i to d in
```
S_{TAG}.

$$c[s] = t_{start}; \forall v \neq s, c[v] = \infty; f[v] = \infty;$$
```
            // c[u] is the arrival time at the node u.
```
$$f[s] = SP[s]; C = \Phi; S = s ;$$
```
            Insert s in priority queue Q.
            while u ≠ d {
                 u = extract_min(Q);
```
$$C = C \bigcup u; S = S - u;$$
```
                 for each node v adjacent to u do {
```
$$if(f[v] > c[u] + d_{uv}(c[u]) + SP_{vd}(c[u])$$
$$c[v] = c[u] + d_{uv}(c[u]);$$
$$f[v] = c[u] + d_{uv}(c[u]) + SP_{vd}(c[u]);$$
```
                     if v is not in Q, insert v in Q;
                 }
                 update Q;
            }
       }
   }
Output the route from s to d.
```

Lemma 6 *SP-TAG_* is correct.*

Proof The A^* algorithm finds the shortest path for a given start time. The heuristic function underestimates the shortest path travel time from the source to the destination since in the computation of the estimate h, the edge travel times are replaced by the minimum values over the entire time horizon. Hence it is not possible for h to exceed the actual travel time. Since the heuristic function is admissible, the A* algorithm finds an optimal solution [24, 37]. Since the heuristic function is monotone, the search process will not open any search node that was once expended and closed. Hence the search is optimal.

Lemma 7 *The time complexity of the SP-TAG_* algorithm is $O(m(\log T + \log n))$ where T is the number of time instants, n is the number of nodes and m is the number of edges in the time aggregated graph.*

Proof   The cost model analysis assumes an adjacency list representation of the graph with two significant modifications. Attached to every adjacent node in the linked list is the travel time series.

For every node extracted from the priority queue Q, there is one edge time series look up and a priority queue update for each of its adjacent nodes. The time complexity of this step is $O(\log T + \log n)$. The asymptotic complexity of the algorithm would be

$$O(\Sigma_{v \in N}[degree(v).(\log T + \log n]) = O(m(\log T + \log n)).$$

3.4 Shortest Path Algorithm for a Given Start Time in a Non-FIFO Network (NF-SP-TAG)

If the travel times do not exhibit the FIFO property it is not guaranteed that an early start at any node ensures an early arrival at any subsequent node. There would be cases where postponing the start at an intermediate node (by waiting) might lead to a reduction in the total travel time. This is illustrated in Fig. 3.7. Selecting the departure time at a node by choosing the earliest availability of the edge, since in some cases, a wait at an intermediate node can lead to a decrease in the total travel time. For example, in Fig. 3.7 a greedy selection of the departure time at node N2 would lead to an arrival at node N3 at $t = 7$, as shown in Fig. 3.7 (ii), resulting in a total travel time of 6 units, which is clearly a non-optimal solution. An optimal solution would be to start at node N1 at $t = 1$ at node $N1$, wait at node N2 for 1 time unit, leading to an arrival at node N3 at $t = 5$, and hence a decrease in total travel time.

In this section, an algorithm that computes a shortest path in a non-FIFO network is described. The key idea behind the algorithm is the insight that by formulating the problem in terms of arrival times at nodes rather than the earliest departure times, we can cast the problem in a transformed problem space, where optimal substructure is valid.

Arrival Time Series Transformation (ATST): The edge weight series in a time aggregated network represents the travel times at various instants. The time series on an edge ij in the transformed network would indicate the arrival time $a_{ij}(t)$ at node j for each departure time t at node i, which would be the sum of departure time t and the travel time $\sigma ij(t)$ $(a_{ij}(t) = t + \sigma ij(t))$.

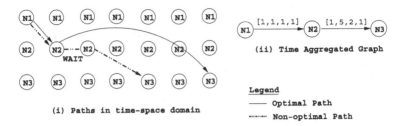

Fig. 3.7 Effect of waits on travel time in non-FIFO networks

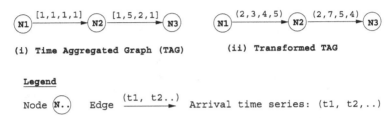

Fig. 3.8 TAG with transformed edge weights

Figure 3.8 illustrates the arrival time series transformation (ATST). The first figure shows a time aggregated graph where the edge weights represent the travel time series. The second figure shows the transformed TAG (T_TAG) where the edge weights are modified to represent the arrival times at the end node of each edge. The algorithm proposed in this section is based on the T_TAG representation.

The algorithm called the NF-SP-TAG computes the shortest path from a given source node to a destination for a give start time, for non-FIFO networks, using the T_TAG representation. In this representation, the edge weights represent the arrival times. Every node has a cost associated with it which represents the arrival time at the node. The algorithm picks the node with the least cost and updates the costs of its adjacent nodes. While computing the costs of the adjacent nodes, the algorithm selects the earliest arrival time that is greater than the departure time (min_arrival operation in the algorithm) at the start node of the edge. The cost of the node is the arrival time at the node. Suppose the cost at the node is $c[u]$ and v is an adjacent node. The edge uv in T_TAG has a time series a_{uv} that indicates the arrival times at v. For every start time t at u. The update on the cost of v is done as $c[v] = c[u] + min_{t \geq c[u]} a_{uv}[t]$. A trace of the algorithm is given in Table 3.5. The table entries are the costs associated with each node (representing the arrival times at the node) at each iteration. The node marked as "closed" is the node with the minimum cost selected for expansion.

Lemma 8 *The NF-SP-TAG algorithm is correct.*

Proof The algorithm runs on the transformed TAG where the edge costs are the arrival times at the end node of the edge. Here we prove that the algorithm computes the shortest path using the greedy strategy. The key idea behind the algorithm is that,

---

Algorithm 3 Shortest Path (NF-SP-TAG) Algorithm

Input:

1) $G(N, E)$: a graph G with a set of nodes N and a set of edges E;

    Each node $n \in N$ has a property:

        Node Presence Time Series : series of positive integers;

    Each edge $e \in E$ has two properties:

        Edge Presence Time Series,

        Arrival_time series : series of positive integers;

      $a_{u,v}(t)$ - arrival time at v for a start time t at u.

2) s: Source node, $s \subseteq N$; 3) d: Destination node, $d \subseteq N$;

4) t_{start}: Start Time;

Output: Shortest Route from s to d for t_{start}

Method:

        $c[s] = t_{start}$; $\forall v \neq s, c[v] = \infty$;

        // $c[u]$ is the cost at the node u.

        Insert s in priority queue Q.

        while Q is not empty do {

            $u = extract\_min(Q)$;

            for each node v adjacent to u do {

                $t = min\_arrival((u, v), c[u])$;

                if $t + \sigma_{u,v}(t) < c[v]$ {

                    $c[v] = t + \sigma_{u,v}(t)$; $parent[v] = u$;

                    if v is not in Q, insert v in Q;

                }

            update Q;

            }

        }

    }

    Output the route from s to d.

---

once the start time is fixed, the earliest arrival at any node implies a shortest path journey. If this is not the case, it contradicts the earliest arrival. The algorithm, at every step, picks the node with the least cost and expands theirs node. While expanding the node, it selects the minimum of all the edge costs, that is greater than the arrival time at the node. The cost of a node is updated as follows using the minimum in the edge time series $c[v] = c[u] + min_{t \geq c[u]} a_{uv}[t]$. $Minimum_{\forall t > t1}(a_{ij}[t]) < Minimum_{\forall t > t2}(a_{ij}[t])$ if $t1 \leq t2$.

Since the algorithm always picks the minimum of the edge costs, it ensures the earliest possible arrival.

Lemma 9 *The time complexity of the NF-SP-TAG algorithm is $O(m(T + \log n))$ where T is the number of time instants, n is the number of nodes and m is the number of edges in the time aggregated graph.*

Proof The cost model analysis assumes an adjacency list representation of the graph with one significant modification. Attached to every adjacent node in the linked list is the arrival time series with the start time instants.

Table 3.5 Trace of the
SP-TAG-nf algorithm for the
network shown in Fig. 3.8

Iteration	N1	N2	N3
1	1 (closed)	∞	∞
2	1	2 (closed)	∞
3	1	2	5 (closed)

For every node extracted from the priority queue Q, there is one edge series look up to find the earliest arrival, and a priority queue update for each of its adjacent nodes. The time complexity of this step is $O(T + \log n)$. The asymptotic complexity of the algorithm would be $O(\Sigma_{v \in N}[degree(v).(T + \log n]) = O(m(T + \log n))$.

3.5 Experimental Analysis

In this section, the experimental analysis of the SP-TAG, SP-TAG* and NF-SP-TAG algorithms are provided. The purpose of the performance evaluation of the algorithm is to compare the run-times with algorithms based on a time-expanded graph.

3.5.1 Experiment Design

Figure 3.9 illustrates the experiment design to compare the performance of the proposed algorithm and the algorithm based on a time expanded network. Time expanded graphs make copies of the original network for every time instant under consideration. The model used for the proposed algorithm is time-aggregated graphs. In our experiments the following were selected as the independent parameters: (1) network size represented by number of nodes; and (2) the length of the time interval in terms of number of time instants. The data sets have two main components: (1) the network data that consists of the graph structure and (2) the travel time series. The networks chosen are road maps from the Minneapolis downtown area with radii of 0.5, 1, 2 and 3 miles. This is appended with travel time series of various lengths. The travel time series were synthetically generated. This data was fed to both a time expanded graph generator, which generates the expanded graph encoding the travel time information. An algorithm for computing the shortest path for a given start time was run on this graph. The algorithms were was run on the same dataset and the results were compared.

The experiments were conducted on a SUN Solaris workstation with 1.77 GHz CPU, 1 GB RAM and UNIX operating system. Each experimental result reported in the following sections is the average over 10 experiment runs with networks generated using the same input parameters, but with different destination nodes.

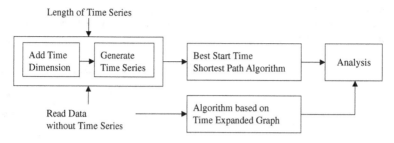

Fig. 3.9 Experiment design

Table 3.6 Description of
datasets

Dataset	Radius (miles)	Number of nodes	Number of edges
1	0.5	111	287
2	1	277	674
3	2	562	1443
4	3	786	2106

3.5.2 Experimental Results and Analysis

Three questions were explored: (1) How does the network size (number of nodes, number of edges) affect the performance of the algorithms? (2) How does the length of the time series affect the performance of the algorithms? (3) How do the two representations, time expanded graph and time aggregated graph, compare with respect to algorithm performance?

Experiment 1: how does the network size affect the performance of the algorithms?

The purpose of the first experiment was to evaluate how the network size in terms of the number of nodes affects the performance of the algorithms. The length of the travel time series was kept constant, and the network size was varied to observe the run times of both fixed start time (SP-TAG) algorithms and time-expanded graph based algorithms. Performance of NF-SF-TAG algorithm was compared to a label correcting algorithm which could handle non-FIFO travel times.

The experiment was done with four datasets that represent the road maps from the Minneapolis downtown area of 0.5, 1, 2 and 3 miles radius. The length of the time series was fixed at 240. The number of nodes and edges in these datasets are provided in Table 3.6. Figure 3.10 shows the run-time of the fixed start time algorithm based on the time aggregated graph and the performance of the algorithm based on the time expanded graph with the A* based algorithm displaying a faster runtime among the TAG algorithms. NF-SP-TAG algorithm displays faster performance compared to label correcting algorithm (Fig. 3.11).

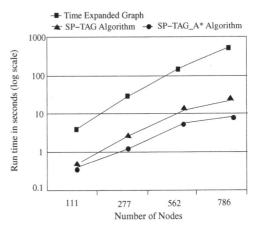

Fig. 3.10 Fixed start time, FIFO algorithm: run-time with respect to network size

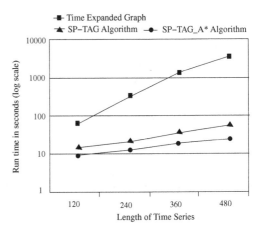

Fig. 3.11 Fixed start time, FIFO algorithm: run-time with respect to time series length

Experiment 2: how does the length of the time series affect the performance of the algorithms?

The second experiment evaluated the effect of the number of time instants on the performance of the algorithms. The length of the time series was varied, while maintaining the network size constant. The number of time instants was varied from 120 to 480 and the network size parameters were fixed at 562 nodes and 1443 edges. As seen in Fig. 3.11, the SP-TAG algorithms perform better, with the SP-TAG* being the fastest.

Experiment 3: how does the edge/node ratio of the network affect the performance of the algorithms?

The third experiment evaluates the effect of edge/node ratio on the performance (Fig. 3.12). The network size was maintained constant while the length of time series was varied. The edge/node ratio was varied from 2 to 6 and the network parameter

Fig. 3.12 NF-SP-TAG: run-time with respect to network size

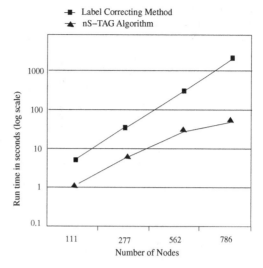

Fig. 3.13 NF-SP-TAG Algorithm: run-time with respect to length of time series

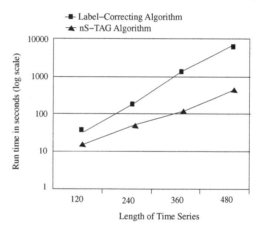

was fixed at 1000 nodes and the number of time instants was fixed at 200 (Fig. 3.13). The networks were generated using SP-RAND network generator [18]. As seen in Fig. 3.14, the SP-TAG algorithm performs better, with the A* based algorithm displaying a faster runtime among the TAG algorithms. As illustrated by Fig. 3.15 TAG based NF-SP-TAG algorithm perform better than time expanded based algorithm.

4: How do the two representations, time expanded graph and time aggregated graph, compare with respect to algorithm performance?

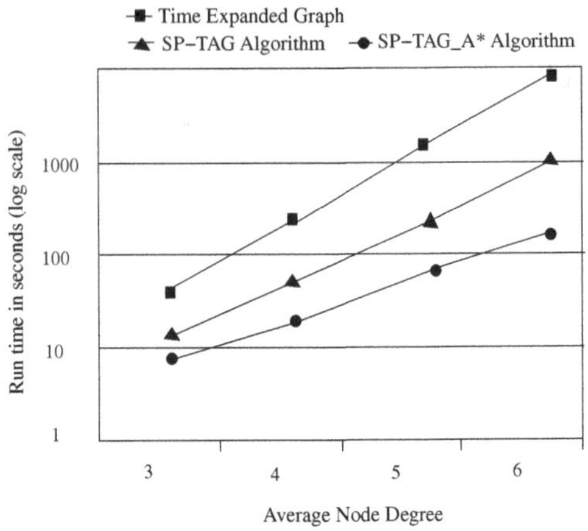

Fig. 3.14 SP-TAG algorithm: run-time with respect to average node degree

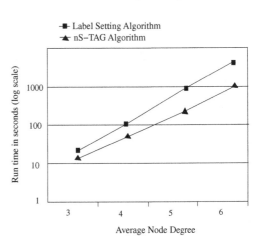

Fig. 3.15 NF-SP-TAG algorithm: run-time with respect to average node degree

Based on the results of Experiments (1), (2), and (3), it can be seen that algorithms based on the time aggregated graph perform better than those based on the time expanded graph.

3.6 Summary

This chapter discusses algorithms for a fixed start time for both FIFO and non-FIFO networks. The formulation of these algorithms is based on a model for spatio-temporal networks called time-aggregated graphs. The study shows that when the start time is restricted and the travel times follow the FIFO property, greedy strategy can be used to formulate the shortest path algorithm. An A* version based on an admissible and monotone heuristic for the same problem, which gives a better performance compared to the greedy algorithm, is also presented.

When the network travel times do not follow FIFO property, it is not possible to apply greedy method to find the shortest route. But it is observed that after a network transformation, a greedy algorithm (NF-SP-TAG) could be formulated. This is based on the fact that when the start time is fixed, the earliest arrival means the least possible travel time, for the given start time.

Chapter 4
Best Start Time Journeys

Abstract The time dependence of parameters in a spatio-temporal network adds to the semantics of common network operations. The result of any analysis in a time dependent network depends on the time at which it is performed. Shortest path from an origin to a destination can vary significantly depending on the start time. This leads to an important and interesting formulation of shortest path computation, "When is the best time to start a journey so that the time spent in the network is minimized?" This chapter describes this formulation in detail and presents algorithms for the computation of 'best start time' shortest paths.

4.1 Introduction

Best start time can be defined as the departure time at the start node that would minimize the time spent in the network. In a time dependent network, the shortest route is time dependent and hence by sometimes postponing the start of the journey could reduce the time spent in the network. This can be true in both FIFO and non-FIFO networks. In addition to the travel time, the routes also can change over time for a given source, destination pair. For example, consider the time dependent graph shown in Fig. 4.1. Here, the shortest path for a start time $t = 1$ is N1–N3–N4–N5 which takes a travel time of 12 units. When the start time is moved to $t = 4$ the shortest path changes to N1–N2–N4–N5 with a travel time of 10 units whereas the route N1–N3–N4–N5 takes 15 time units, illustrating a change in route as well as travel time, as the start time changes. Hence, finding the best start time involves the minimization of travel time over the time and space domains. The formulation of algorithms to compute the paths that take the least commute time becomes non-trivial since most of the techniques that are used in static networks might not be applicable in dynamic scenarios. Since the network changes in its parameter values and the topology, meeting the requirements of efficiency and correctness can pose challenges. The potential waits at intermediate nodes can increase the total

B. George and S. Kim, *Spatio-temporal Networks*,
SpringerBriefs in Computer Science, DOI: 10.1007/978-1-4614-4918-8_4,

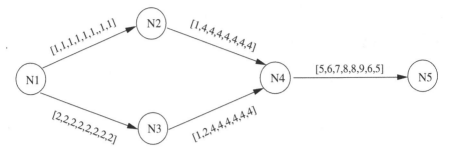

Fig. 4.1 Illustration of route change with start time

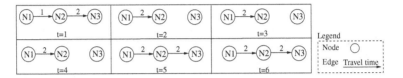

Fig. 4.2 Network at various instants

journey time even if an initial part of the path turns out to be optimal. Figure 4.2 shows a spatial network that changes with time. The figure shows the snapshots of the network at various instants of time, and the edges are marked with the travel times. It is significant to note that the prefix journeys of the best start time shortest path journey are not always optimal since some optimal prefix journeys can lead to longer waits at intermediate nodes. The best start time for a journey from node N1 to N3 is $t = 4$, which takes 4 time units. The optimal path from N1 to N3 that starts at $t = 4$ is not optimal for the intermediate node N2. The best start time for a path from N1 to N2 is $t = 1$, which proves to be sub-optimal for a journey from N1 to N3. The lack of an optimal substructure in the best start time shortest paths rules out the possibility of using a greedy strategy in the algorithm design.

This chapter presents algorithms to compute shortest paths for the best start time and consequently the least commute time paths for both FIFO and non-FIFO networks (TI-SP-TAG*, CP-NF-BEST, and BEST algorithms respectively). The BEST algorithm for non-FIFO networks uses a node cost time series instead of a scalar node cost. The entries in the time series are updated when a path of smaller cost is found. The algorithm iterates until every entry reaches a minimum value and hence does not depend on the greedy choice property. Concurrent Prioritized Non-FIFO BEst STart (CP-NF-BEST) Algorithm is a logically concurrent version of NF-SP-TAG algorithm.

The algorithm to find the best start time path in a FIFO network essentially iterates the SP-TAG* algorithm, described in Sect. 3.3.2, for every time instant. Since the edge costs follow the FIFO property, the path computed for each step is correct and the start time instant that results in the least travel time would be the best start time. One key step in this algorithm is the computation of the estimated cost at each node and the algorithm tries to reuse computed costs from the previous iteration to reduce computational cost.

4.2 Basic Concepts

Spatial networks that show time-dependence serve as the underlying networks for many applications such as routing in transportation networks. Traditionally graphs have been extensively used to model spatial networks (e.g. road networks) [40]; weights assigned to nodes and edges are used to encode additional information. In a real world scenario, it is not uncommon for these network parameters to be time-dependent. It is important to be able to formulate computationally efficient and correct algorithms for the shortest path computation that take into account the dynamic nature of the networks. Models of these networks need to capture the possible changes in topology and values of network parameters with time and provide the basis for the formulation of computationally efficient and correct algorithms for the frequent computations like shortest paths.

Given a set of frequent queries posed by an application on a spatial network and the pattern of variations of the spatial network with time, we need to find a model that supports efficient and correct algorithms for computing the query results, while trying to minimize the storage and cost of computation. In this section we discuss the basics of the model used to represent time dependent spatial networks called "Time Aggregated Networks" [16]. The algorithms presented in this paper are formulated based on this model. Time aggregated graphs can not only capture the time-dependence of network parameters, but also account for the possibility of edges and nodes being absent during certain instants of time.

4.2.1 The Conceptual Model

A graph $G = (N, E)$ consists of a finite set of nodes N and edges E between the nodes in N. If the pair of nodes that determines the edge is ordered, the graph is directed; if it is not, the graph is undirected. In most cases, additional information is attached to the nodes and edges. In this section, we discuss how the time dependence of these edge/node parameters are handled in the proposed time-aggregated graph model.

We define the time-aggregated graph as follows.

$taG = (N, E, TF, f_1 \ldots f_k, g_1 \ldots g_l, w_1 \ldots w_p | f_i : N \rightarrow \mathbb{R}^{TF}; g_i : E \rightarrow \mathbb{R}^{TF}; w_i : E \rightarrow \mathbb{R}^{TF})$

where N is the set of nodes, E is the set of edges, TF is the length of the entire time interval, $f_1 \ldots f_k$ are the mappings from nodes to the time-series associated with the

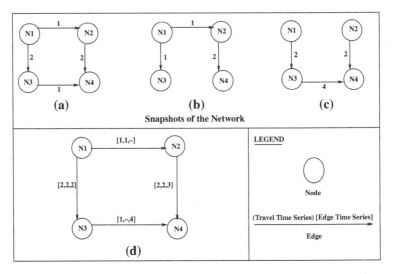

Fig. 4.3 Network at various time instants and the time aggregated graph. Snapshots of a network at time instants **a** t = 1, **b** t = 2, **c** t = 3. **d** Time aggregated graph

nodes, $g_1 \ldots g_l$ are mappings from edges to the time series associated with the edges, and $w_1 \ldots w_p$ indicate the time dependent weights (e.g. travel times) on the edges.

Each edge has an attribute, called an edge time series that represents the time instants for which the edge is present. This enables the time aggregated graph to model the topological changes of the network with time. We assume that each edge travel time has a positive minimum and the presence of an edge at time instant t is valid for the closed interval $[t, t + \sigma]$.

Figure 4.3a, b, c shows a network at five time instants. The network topology and parameters change over time. For example, edge N3–N4 is present at time instants $t = 1, 3$, and absent at $t = 2$. The time aggregated graph that represents this dynamic network is shown in Fig. 4.3d. In this figure, edge N3–N4 has an attribute, $[1, -, 4]$, which is its weight time series, indicating the weight of the edge at instants $t = 1, 2, 3$. Though this model can include spatial properties at nodes and edges, these properties are not incorporated in the algorithms discussed. Figure 4.4a shows the time aggregated graph (corresponding to Fig. 4.3a, b, c) and a time expanded graph that represents the same scenario. Edge weights in a time expanded graph are not explicitly shown as edge attributes; instead they are represented by edges that connect the copies of the nodes at various time instants. For example, the weight 1 of edge N1–N2 at $t = 1$ is represented by connecting the copy of node N1 at $t = 1$ to the copy of node N2 at time $t = 2$. The time expansion for the example network needs to go through 7 steps since the latest edge traversal in the network ends at $t = 7$. The traversal of the edge N3–N4 that starts at $t = 3$ ends at $t = 7$, the travel time of the edge being 4 units. The number of nodes is larger by a factor of T, where T is the number of time instants and the number of edges is also larger

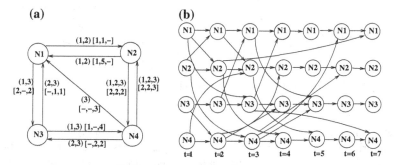

Fig. 4.4 Time-aggregated graph (**a**) versus time expanded graph (**b**)

in number compared to the time-aggregated graph. If the value of T is very large in a spatial network, it would result in enormously large time expanded networks and consequently slow computations.

Comparison of Storage Costs with Time Expanded Networks

According to the analysis in [41], the memory requirement for time expanded network is $O(nT) + O((n + m)T)$, where n is the number of nodes and m is the number of edges in the original graph. The framework of a time aggregated graph would require a memory of $O(n + m)$, where n is the number of nodes and m is the number of edges. Edges and nodes with time-varying attributes have attribute time series associated with them. If the average length of the time series is $\alpha(\leq T)$, the memory required is $O(\alpha m + \alpha n)$, assuming an adjacency list representation. The total memory requirement for a time aggregated graph is $O(n + m + \alpha m + \alpha n)$. This comparison shows that the memory usage of time-aggregated graphs is less than that of time expanded graphs if $\alpha < T$.

4.2.2 Basic Design Space of Shortest Path Algorithms

One of the most frequent queries on any spatio-temporal network is the computation of shortest paths. In time dependent networks, the shortest path and the traversal time are dependent on the start time. For example, a shortest path from node N1 to N5 in Figure 4.3 for the start time $t = 1$ is N1–N3–N4–N5 with a travel time of six units. If the start is postponed to $t = 3$, the shortest path changes to N1–N2–N4–N5 and the travel time drops to four units. Due to the time dependence of shortest paths, in a spatio-temporal network it is possible to raise interesting queries such as "When is the best time to start a journey so that time spent in the network is the least?". Sections 4.3 and 4.4 describe the formulations of two shortest path problems, first for a fixed start time and second, for the least travel time. The design of these algorithms effectively utilizes certain properties of the time dependent parameters (such as the FIFO property

Fig. 4.5 Classification of TAG-based shortest path algorithms

of travel time). The classification of the algorithms is shown in Fig. 4.5. The shortest path algorithms show significantly different properties based on their formulations. For example, shortest path computation for a fixed start time might display optimal prefix property under certain assumption on travel time characteristics. Computation of best start time shortest path under non-FIFO travel times might not always display optimal prefix property ruling out popular design techniques such as greedy and A* based algorithms. Based on these characteristics, shortest path computation on time aggregated graph falls under various categories as shown in Fig. 4.5.

4.2.3 Algorithmic Challenges

A time dependent graph might not display some properties that would make some common algorithm design techniques such as dynamic programming and greedy strategy feasible. For example most time dependent graphs do not exhibit optimal prefix property, thus making it impossible to apply greedy methods in shortest path computations. Figure 4.6 shows a time dependent network represented as a time aggregated graph. If the edge costs are assumed to be edge travel times, the cost of a node indicates the arrival time at the node. When the least cost node is expanded, the costs of the outgoing edges are chosen as the costs at the arrival time. In cases where an edge is not available, the cost at the earliest available time is selected. Table 4.1 shows the execution trace of Dijkstra's algorithm on the time aggregated graph shown in Fig. 4.6. The schedule in Table 4.2 shows the start time and arrival at each node as computed by the algorithm. As Table 4.2 illustrates, the algorithm computes the route N1–N2–N4–N5 as the shortest path for the start time $t = 1$ at node N1. The total travel time for the route is 14 time units. It can be seen that there is another route N1–N3–N4–N5 for the same start time which takes a travel time of 12 time units. This is an optimal route from N1 to N5, which suggests that the algorithm might not always compute the shortest path.

Shortest Path Computation and Stationarity

Stationarity means the following: if two reward sequences $R_1, R_2, \ldots$ and $S_1, S_2, \ldots$ begin with the same reward then the sequences should be preference ordered the same way as the sequences $R_2, R_3, \ldots$ and $S_2, S_3, \ldots$ [2, 39]. In shortest route computation, if two journeys $e1, e2, e3, \ldots$ and $f1, f2, f3 \ldots$ start at the same node, then the preference order of the journeys should not change for another start time. In the time dependent graph shown in Fig. 4.6 the shortest path for a start time $t = 1$ is N1–N3–N4–N5 which takes a travel time of 12 units.

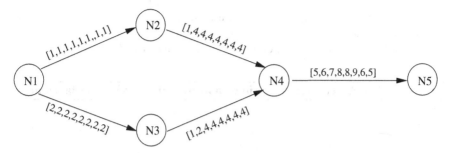

Fig. 4.6 Illustration of shortest path computation

Table 4.1 Trace of Dijkstra's algorithm for the network shown in Fig. 4.6

Iteration	N1	N2	N3	N4	N5
1	1 (closed)	∞	∞	∞	∞
2	1	2 (closed)	3	∞	∞
3	1	2	3 (closed)	6	∞
4	1	2	3	6 (closed)	15
5	1	2	3	6	15 (closed)

Table 4.2 Start and arrival times at nodes

Route computed by algorithm					Optimal route					
Node	N1	N2	N3	N4	N5	Node	N1	N2	N3	N4	N5
Arrival time	1	2	-	6	15	Arrival time	1	-	3	7	13
Start time	1	2	-	6	15	Start time	1	-	3	7	13
Total travel time = 14						Total travel time = 12					

When the start time is moved to $t = 4$ the shortest path changes to N1–N2–N4–N5 with a travel time of 10 units whereas the route N1–N3–N4–N5 takes 15 time units, illustrating a change in preference order as the start time changes, and hence displaying non-stationarity. The lack of stationarity would eliminate the possibility of using dynamic programming as a design technique.

4.3 Time Iterative SP-TAG* (TI_SP-TAG*) Algorithm for FIFO Networks

Under the assumption of FIFO travel times, the best start algorithm can be formulated as an iterative formulation of SP-TAG* search algorithm (Time Iterative SP-TAG*). The SP-TAG* version was chosen instead of the greedy SP-TAG algorithm since it performed better in the fixed start time algorithms. A comparative experimental analysis is provided to evaluate this decision in Sect. 4.5.

The evaluation function $f(n)$ of a node n is formulated as $f(n) = g(n) + h(n)$.

Algorithm 1 A* based Best Start Time Shortest Path (BEST_A*) Algorithm

Input:
```
    1) G(N, E): a graph G with a set of nodes N and a
       set of edge s E;
       Each node n ∈ N has a property:
           Node Presence Time Series : series of positive
integers;
           Each edge e ∈ E has two properties:
           Edge Presence Time Series,
           Travel_time series : series of positive integers;
           σ_{u,v}(t) - travel time of edge uv at time t.
    2) s: Source node, s ⊆ N; 3) d: Destination node, d ⊆ N;
    4) t_{start}: Start Time;
```
Output: Shortest Route from s to d for t_{start}

Method:
```
       Preprocess: find the shortest path from node i to d in
S_{TAG}.
       for (i = 1toT) {
           t_{start} = i;
           c[s] = t_{start}; ∀v ≠ s, c[v] = ∞; f[v] = ∞;           // c[u] is the
           arrival time at the node u.
           f[s] = SP[s]; C = Φ; S = s ;
           Insert s in priority queue Q.
           while u ≠ d {
               u = extract_min(Q);
               C = C ⋃ u; S = S − u;
               for each node v adjacent to u do {
                   if(f[v] > c[v]
                       c[v] = c[u] + d_{uv}(c[u]);
                       f[v] = c[u] + d_{uv}(c[u]) + SP_{vd}(c[u]);
                       if v is not in Q, insert v in Q;
                   }
               update Q;
               }
           }
       Compute the travel time as the difference(start time,
arrival time);
           }
       }
   Output the route from s to d.
```

$g(n)$ is the actual cost to reach the node n from the start node s, which is the time taken to reach the current node from the start node. $h(n)$ is the estimated cost from the node n to a destination node d. We propose $h(n)$ to be the shortest path travel time from node n to the destination node d computed based the least travel time on each edge. This cost function is used to compute the shortest path for every start time and the least travel time is chosen from the shortest path travel times for all start times. The algorithm runs the A* algorithm for every start time and computes the shortest path for each start time. Since the heuristic function is admissible and monotone (See Lemma 4 and Lemma 5), the computed shortest paths are correct and the search is optimal. Due to the FIFO property of the travel times, the best start time would be the start time that yields the least travel time among the shortest paths over all start times.

Performance Tuning: It is observed that in a FIFO network, it is possible to take advantage of the results obtained from the previous steps in the iteration on time instants. When computing the shortest path for the start time t, the shortest path for the start time $t - 1$ has already been computed. In a FIFO network, the earliest arrival at the destination node (d) for a given start time at the source node (s) implies the least travel time for the same start time. This leads to the observation that an admissible estimate of the cost from a node i on the shortest path from node s to node d can be computed from the costs computed in the previous iteration as follows. The admissible estimate of travel time from node i to node d is $(t - 1) + d_{si}(t - 1) + SP_{id}(t - 1 + d_{si}(t - 1)) - (t + d_{si}(t))$ where $d_{si}(t)$ is the shortest travel time from s to node i for start time t, $SP_{id}(t)$ is the shortest travel time from i to destination d for start time t. $(t - 1) + d_{si}(t - 1)$ is the arrival time at node i for start time $t - 1$ and hence $(t - 1) + d_{si}(t - 1) + SP_{id}(t - 1 + d_{si}(t - 1))$ is the earliest arrival time at the destination d for the start time $t - 1$. The estimate is computed by subtracting the earliest arrival time at i for the (next) start time instant t ((ie.) $t + d_{si}(t)$). The fact that this is admissible can be proved as follows:

Since the network is FIFO, the arrival time for start time t cannot be earlier than that for $t - 1$. Or,

$t + d_{si}(t) \geq (t - 1) + d_{si}(t - 1).$

$t + d_{si}(t) + SP_{id}(t + d_{si}(t)) \geq (t - 1) + d_{si}(t - 1) + SP_{id}(t - 1 + d_{si}(t - 1))$

This uses the property that the arrival at the destination d for start time t cannot be earlier than that for a start at $t - 1$.

Therefore, $SP_{id}(t + d_{si}(t)) \geq (t - 1) + d_{si}(t - 1) + SP_{id}(t - 1 + d_{si}(t - 1)) - (t + d_{si}(t))$ and hence $(t - 1) + d_{si}(t - 1) + SP_{id}(t - 1 + d_{si}(t - 1)) - (t + d_{si}(t))$ is an admissible estimate for the shortest path from an intermediate node i to d. This estimate can be used only in the case of nodes that are a part of the shortest path at the earlier iteration. If this is not the case, we use the heuristic based on static TAG (S_TAG) as explained in Sect. 3.3.2 to estimate the cost.

4.4 Best Start Time Shortest Path Algorithms for Non-FIFO Networks

A path that takes the smallest travel time for a source-destination traversal over the entire time horizon (called 'Best Start Time shortest Path') can be computed. This is significant since it suggests that it is possible to reduce the travel time for the same source-destination pair if the travel starts at the "right" time instant. The formulation of algorithms to compute the paths that take the least commute time becomes non-trivial since most of the techniques that are used in static networks might not be applicable in dynamic scenarios.

This chapter presents algorithms to compute shortest paths for the best start time and consequently the least commute time paths for both FIFO and non-FIFO graphs (TI-SP-TAG*, CP-NF-BEST, and BEST algorithms respectively). The BEST algorithm for non-stationary networks uses a node cost time series instead of a scalar node cost. The entries in the time series are updated when a path of smaller cost is found. The algorithm iterates until every entry reaches a minimum value and hence does not depend on the greedy choice property. We also propose a logically concurrent version of NF-SP-TAG algorithm (CP-NF-SP-TAG)for the non-FIFO networks and compare its performance with the BEST algorithm which uses a label-correcting strategy [5].

4.4.1 Best Start Time Shortest Path (BEST) Algorithm (Label Correcting Approach)

The algorithm presented in this section uses the time aggregated graph to model a time dependent spatial network. While computing the best start time, each node needs to keep track of the travel times to the destination for every start time instant. The proposed algorithm attributes each node with a time series, with ith entry representing the current, least travel time to the destination node for the start time t_i. Due to the lack of optimality of prefix paths and lack of ordering of nodes based on the costs (ie. travel times), nodes cannot be selected and "closed" based on a minimum scalar cost. The algorithm uses an iterative, label correcting approach [4] and each entry in a node time series is modified according to the following condition.

$$C_u[t] = minimum\{C_u[t], \sigma_{uv}(t) + C_v[t + \sigma_{uv}(t)]\} \qquad (4.1)$$

where,

$uv \in E$
$C_u[t]$—travel time from $u \in N$ to the destination for the start time t
$\sigma_{uv}(t)$—travel time of the edge uv at time t

The algorithm maintains a list of all nodes that change its cost according to the condition and terminates when there is no further improvement indicated by an empty list. Though the list can be implemented using several data structures, studies on static networks [4, 49] have shown that the Two_Q implementation [35] of label correcting algorithms performs the best on road networks.

The search starts at the destination node and proceeds to update the remaining nodes, finally finding the best start time shortest paths from all nodes to the destination. Figure 4.7 illustrates the trace of the algorithm on a small network. In this example, the destination node is the node N4. The node cost series C_4 is initialized to $[0, 0, 0, 0, 0]$ and the cost series C_i, $i = 1, 2, 3$ are initialized to $[\infty, \infty, \infty, \infty, \infty]$. The nodes that have N4 in their adjacency lists (that is, all nodes N_i such that $N_i N4 \in E$), N2 and N3 are relaxed according to condition (4.1). These nodes are added to the queue since there is a change in their cost series. The steps continue until the queue is empty, indicating that there is no further cost improvement at any of the nodes. At every iteration, the node that contributes to a cost improvement is stored in a descendant array to facilitate the trace of the shortest paths when the algorithm terminates. At the termination, the cost time series has the travel times for every start time $t = 1, 2, \ldots T$. For example, the cost time series of node N1 shows that the travel times from N1 to N4 for start times $t = 1$ is 4 time units, while the best start time at this node is $t = 4$, which results in a travel time of 2 time units and a best start time shortest path N1–N2–N4. N1–N2 takes 1 time unit at $t = 4$, reaches N2 at $t = 5$ and continues on N2–N4 at $t = 5$, reaching N4 at $t = 6$, taking a total travel time of 2 time units. A more detailed trace is shown in Table 4.3.

Algorithm 2 BEST Algorithm

Input:
   $G(N, E)$: a graph G with a set of nodes N and a
   set of edges E;
       Each node $n \in N$ has a property:
           Node Presence Time Series : series of positive
integers;
       Each edge $e \in E$ has two properties:
           Edge Presence Time Series,
           Travel_time series : series of positive integers;
           $\sigma_{u,v}(t)$ - travel time of edge uv at time t.
   Output:
   Best Start Time shortest route from s to d;
   Initialize;
   While Queue not Empty
       v = Dequeue();
       For every node u such that $uv \in E$
           For every entry in the cost series C_u of u
               if $C_u(t) > \sigma_{uv}(t) + C_v(t + \sigma_{uv}(t))$
                   Update $C_u(t)$;
                   Enqueue(u);
                   Update the descendant array of u.
   Find the minimum entry in the node time series.
   Return the BestStartTime and the ShortestRoute;

Lemma 10 *The algorithm terminates and computes the best start time paths from every node to the destination.*

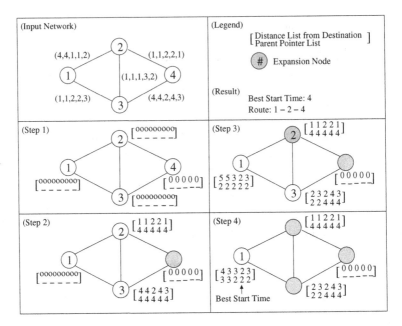

Fig. 4.7 Trace of the BEST algorithm

Table 4.3 Trace of the BEST algorithm for the network shown in Fig. 4.7

Iteration	N1	N2	N3	N4	Queue
1	$\infty\ldots\infty$	$\infty\ldots\infty$	$\infty\ldots\infty$	[0, 0, 0, 0, 0]	$N1$
2	$\infty\ldots\infty$	[1, 1, 2, 2, 1]	[4, 4, 2, 4, 3]	[0, 0, 0, 0, 0]	$N2, N3$
3	$\infty\ldots\infty$	[1, 1, 2, 2, 1]	[2, 3, 2, 4, 3]	[0, 0, 0, 0, 0]	$N3$
4	[4, 3, 3, 2, 3]	[1, 1, 2, 2, 1]	[2, 3, 2, 4, 3]	[0, 0, 0, 0, 0]	$N1$
5	[4, 3, 3, 2, 3]	[1, 1, 2, 2, 1]	[2, 3, 2, 4, 3]	[0, 0, 0, 0, 0]	–

Proof The algorithm terminates because there is a positive minimum for the travel time over every path, for every pair of nodes in the network since the edge weights (travel times) are positive and each such path has a finite number of edges. The updates on the costs according to condition (4.1) will generate the optimal travel times from a node to the destination at the termination of the algorithm. This can be proved by induction on the number of edges on the path. The base condition would be for paths with two edges, say from any node u to the destination node d. Every path with two edges from u to d will transit to some node v and then traverse the edge to d which takes the least time. If we assume the inductive hypotheses is true for every path with k edges, the minimality must hold for a path from u with $(k+1)$ edges since we can reach node u that with a minimal k-edge path and append uv with travel time $\sigma_{uv}(t)$.

Lemma 11 *The computational complexity of the BEST algorithm is $O(n^2mT)$, where n is the number of nodes, m is the number of edges and T is the length of the time series.*

Proof The worst case computational complexity of the label correcting algorithm based on Two-Q data structure is $O(n^2m)$ when the node costs and edge weights are scalar quantities [4]. In the BEST algorithm, the relaxation step operates on a time series (node cost and edge weight) of length T. Hence the computational complexity of the algorithm is $O(n^2mT)$.

4.4.2 Best Start Time Algorithm Using ATST (CP-NF-BEST Algorithm)

This section describes a best start time algorithm where the benefits of arrival time series transformation are utilized. The basic idea behind the algorithm is iterating the NF-SP-TAG algorithm, but with logical concurrency. This concurrency is achieved by keeping an open list of nodes such that every node has a copy for every start time. While selecting the node for expansion, the copy of the node with the minimum cost is selected. The pseudocode is provided in Algorithm 6.

Lemma 12 *CP-NF-BEST algorithm is correct.*

Proof The proof follows from the correctness of NF-SP-TAG algorithm (Lemma 8). The algorithm CP-NF-BEST essentially iterates NF-SP-TAG over the entire time period, maintaining a list of copies of open nodes. At each start time (an iteration), the algorithm computes the earliest arrival, which is the shortest duration journey for the given start time. Since there is a minimum for every start time and the algorithm picks the minimum of these durations, the algorithm computes the least duration journey shortest path.

Lemma 13 *The computational complexity of CP-NF-BEST is $O(mT(T + \log n))$ where m is the number of edges, n is the number of nodes, and T is the length of time period.*

Proof The complexity of NF-SP-TAG algorithm is $O(m(T + \log n))$. So, the worst case complexity of CP-NF-BEST is $T.O(mT(T + \log n))$, *i.e.*, $O(mT(T + \log n))$.

4.5 Experimental Analysis

In this section, the experimental analysis of the BEST algorithm and the SP-TAG algorithm are provided. The purpose of the performance evaluation of the algorithm is to compare the run-times with algorithms based on a time-expanded graph.

Algorithm 3 Shortest Path (CP-NF-BEST) Algorithm

Input:
   1) $G(N, E)$: a graph G with a set of nodes N and a
     set of edges E;
     Each node $n \in N$ has a property:
       Node Presence Time Series : series of positive
integers;
     Each edge $e \in E$ has two properties:
       Edge Presence Time Series,
       Arrival_time series : series of positive integers;
      $a_{u,v}(t)$ - arrival time at v for a start time t at u.
   2) s: Source node, $s \subseteq N$; 3) d: Destination node, $d \subseteq N$;
Output: Shortest Route from s to d.
Method:

```
        for (i=1 to T) do
            s_i = i;  ∀v ≠ s, c[v] = ∞;
            // c[u] is the cost at the node u.
            Insert s_i in priority queue Q.
            while Q is not empty do {
                u_i = extract_min(Q);
                if u = d break;
                for each node v adjacent to u do {
                    t = min_arrival((u, v), c[u_i]);
                    if t + σ_u,v(t) < c[v_i] {
                        v_i = t + σ_u,v(t);  parent[v_i] = u;
                        if v_i is not in Q, insert v_i in Q;
                    }
                    update Q;
                }
            }
        }
    Output the route from s to d.
```

4.5.1 Experiment Design

Figure 4.8 illustrates the experiment design to compare the performance of the proposed algorithm and the algorithm based on a time expanded network. Time expanded graphs make copies of the original network for every time instant under consideration. The model used for the proposed algorithm is time-aggregated graphs. In our experiments the following were selected as the independent parameters: (1) network size represented by number of nodes; and (2) the length of the time interval in terms of number of time instants. The data sets have two main components: (1) the network data that consists of the graph structure and (2) the travel time series. The networks chosen are road maps from the Minneapolis downtown area with radii of 0.5, 1, 2 and 3 miles. This is appended with travel time series of various lengths. The travel time series were synthetically generated. This data was fed to both a time expanded graph generator, which generates the expanded

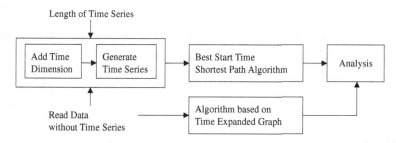

Fig. 4.8 Experiment design

graph encoding the travel time information. An algorithm for computing the shortest path for a best start time was run on this graph. The results were compared to the results from the BEST and TI-SP-TAG* (for FIFO travel times) algorithms.

The experiments were conducted on a SUN Solaris workstation with 1.77 GHz CPU, 1GB RAM and UNIX operating system. Each experimental result reported in the following sections is the average over five experiment runs with networks generated using the same input parameters, but with different destination nodes.

4.5.2 Experimental Results and Analysis

Four questions were explored: (1) How does the network size (number of nodes, number of edges) affect the performance of the algorithms? (2) How does the length of the time series affect the performance of the algorithms? (3) How does the network structure in terms of the edge/node ratio affect the performance? (4) How do the two representations, time expanded graph and time aggregated graph, compare with respect to algorithm performance?

Experiment 1: How does the network size and time series length affect the performance of the BEST algorithm?

The purpose of the first experiment was to evaluate how the network size and the time series length affect the performance of the BEST algorithm. To evaluate the scalability with respect to the network size, the length of the travel time series was maintained constant, and the network size was varied to observe the run times best start time (BEST) algorithms and time-expanded graph based algorithms. The experiment to study the effect of time series length was performed with a fixed network size and varying time series lengths.

The experiment was done with four datasets that represent the road maps from the Minneapolis downtown area of 0.5, 1, 2 and 3 mile radius. The length of the time series was fixed at 240. The number of nodes and edges in these datasets are provided in Table 4.4. Figure 4.9 shows the run-time of the best start time algorithms based on the time aggregated graph and the performance of the algorithm based on

Table 4.4 Description of datasets

Dataset	Radius (miles)	Number of nodes	Number of edges
1	0.5	111	287
2	1	277	674
3	2	562	1443
4	3	786	2106

Fig. 4.9 BEST, CP-NF-BEST algorithms: run-time with respect to network size

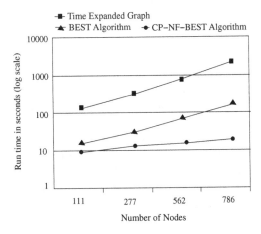

Fig. 4.10 BEST, CP-NF-BEST algorithms: run-time with respect to time series length

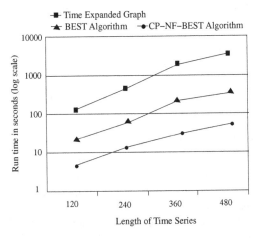

the time expanded graph. The BEST algorithm runs faster than the time-expanded graph based algorithm in all cases; further, its run-time seems to increase at a slower rate.

Experiment 2: How does TI_SP-TAG perform with respect to the network size and the length of the time series?

Fig. 4.11 TI-SP-TAG* algorithm: run-time with respect to network size

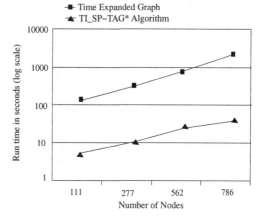

Fig. 4.12 TI-SP-TAG* algorithm: run-time with respect to length of time series

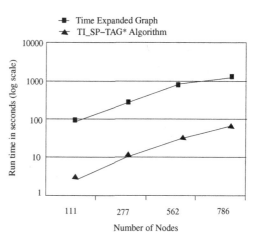

In the second experiment, the performance of TI SP-TAG algorithm was compared to an algorithm that runs on time expanded graph. In the evaluation with respect to series length, the network size was held constant while varying the length of the time series and run-times were observed. The number of time instants was varied from 120 to 480 and the network size parameters were fixed at 562 nodes and 1443 edges. In the other case, the length of the time series was held constant and the network sizes were varied. Figure 4.11 shows the run-time of the fixed start time algorithms based on the time aggregated graph and the performance of the algorithm based on the time expanded graph. As seen in Fig. 4.12, the TI_SP-TAG algorithm performs better, compared to the time expanded graph version. As the length of the time series increases, the number of copies of the entire network required in the case of the time expanded graph increases, resulting in a considerable increase in the size of the entire network, leading to almost exponential increases in run time.

Fig. 4.13 TI_SP-TAG* algorithm: run-time with respect to
average degree of the network

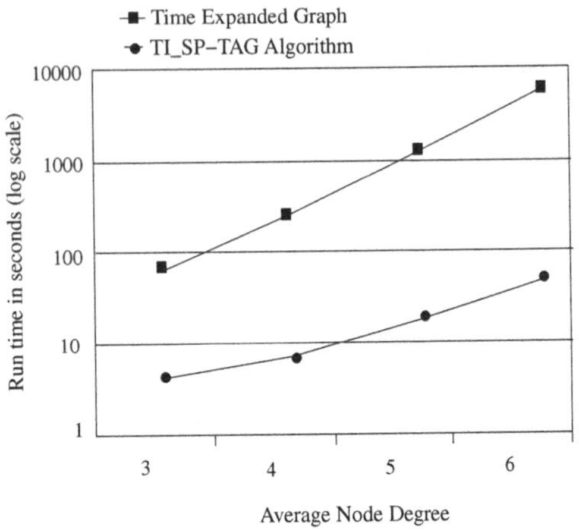

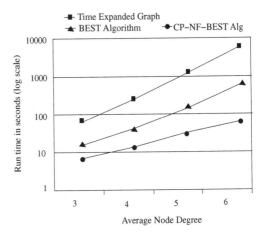

Fig. 4.14 BEST, CP-NF-
BEST algorithms: run-time
with respect to average degree
of the network

Experiment 3: How does the edge/node ratio of the network affect the performance
of the algorithms?

In the third experiment, the effect of edge/node ratio of the network on the performance of the algorithms was evaluated. The network size, and the length of the
time series were held constant and the average degree of the network was varied and
the run-times were observed. The edge/node ratio was varied from 1.5 to 5.5 and the
network parameter was fixed at 1000 nodes and the number of time instants was fixed
at 200. The networks were generated using SP-RAND network generator. As seen in
Fig. 4.13, the TI_SP-TAG* algorithm performs better. Figure 4.14 the performance
of the BEST algorithm and that of the time expanded graph algorithm.

Experiment 4: How does the iterative version of Greedy SP-TAG algorithm compare to the iterative version of SP-TAG* algorithm (TI_SP-TAG)?

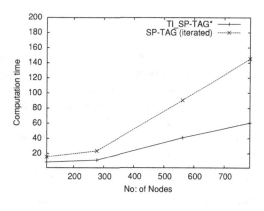

Fig. 4.15 Comparison of TI_SP-TAG* with iterated SP-TAG: run-time with respect to network size

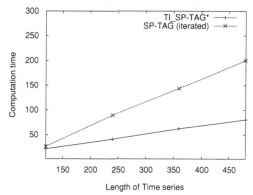

Fig. 4.16 Comparison of TI_SP-TAG* with iterated SP-TAG: run-time with respect to time series length

In the fourth experiment, the iterative version of Greedy SP-TAG algorithm was compared to the iterative version of SP-TAG* algorithm (TI SP-TAG), with respect to the (i) network size and (ii) time series length. In case (i), the length of the time series was kept constant and the network size was varied, whereas in the second case the length of the time series was changed maintaining the network size constant. As seen in Figs. 4.15 and 4.16, the TI_SP-TAG* algorithm performs better than the iterative version of the greedy SP-TAG algorithm.

Experiment 5: How do the two representations, time expanded graph and time aggregated graph, compare with respect to algorithm performance?

Based on the results of Experiments (1) and (2), it can be seen that algorithms based on the time aggregated graph perform better than those based on the time expanded graph. Under the assumption of FIFO travel times, the A* based algorithm based on an admissible, monotone heuristic shows the best performance among the three algorithms.

4.6 Summary

This chapter discusses the flexible start time algorithms for both FIFO and non-FIFO networks. Flexible start time algorithms have significant applications in daily commutes and in logistical services such as freight delivery. This algorithm enables us to find the start time such that the travel time is minimized over the entire time horizon and hence is relevant in the context of fuel consumption.

The Best Start Time algorithm uses a node cost time series instead of a scalar node cost. The entries in the time series are updated when a path of smaller cost is found. The algorithm iterates until every entry reaches a minimum value and hence does not depend on the greedy choice property. This removes the FIFO restriction from the edge travel times. We also present the experimental analysis of the best start time algorithm for both FIFO and non-FIFO networks.

Chapter 5
Spatio-temporal Network Application

Abstract This chapter provides brief descriptions of key real world domains where spatio-temporal networks play a significant role such as multimodal transportation networks and sensor networks. The chapter illustrates the modeling of multimodal transportation networks and sensor networks using time aggregated graphs.

5.1 Multimodal Transportation Networks

Multimodal transportation network can be viewed as an integrated network that consists of multiple component networks that often belong to various modes of transportation. For example a public transportation system that serves a metropolitan area can consist of bus networks, subway train networks, and ferry systems, each of which would typically consist of multiple routes and trips. These networks interact with one another through facilities for inter-mode transfers, while also allowing transfers across various routes within a single mode network. A significant feature of most multimodal transportation networks is that they are schedule-based. The schedule-based operation of the services make such networks time-dependent and computatations of routes need to account for the time-dependence, which requires a model that can capture the temporal dimension of the network.

5.1.1 Modeling Multimodal Networks

Multimodal networks display time dependence in the availability of services (through schedules) and in network traversal costs (such as travel times) which can depend on the congestion levels. Since routing in a multimodal network needs to account for its time variant nature, there is a need for an efficient model that can represent the schedules and the time variant network traversal costs. Figure 5.1a shows

B. George and S. Kim, *Spatio-temporal Networks*,
SpringerBriefs in Computer Science, DOI: 10.1007/978-1-4614-4918-8_5,
© The Author(s) 2013

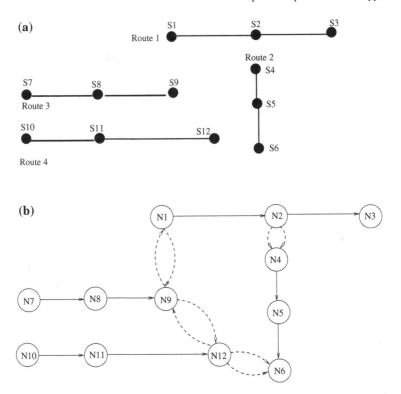

Fig. 5.1 An example multimodal network

a simple multimodal network that consists of four routes in total. Each stop has an arrival-departure schedule associated with it. Figure 5.1b shows the network representation of the multimodal network. Each stop is represented as a node and the connectivity across stops are represented as network links and the cost associated with a link is assumed to be the travel time along the link. In addition to the links that connect the stops along a route, there are links that connect across routes or modes. These indicate the facility to transfer from one route to another and can be created based on factors such as proximity between the stops. Such links are represented using broken lines in the figure, an example being the link $N2$–$N4$, which represents the facility to transfer from Route 1 to Route 4.

For the sake of simplicity it is assumed that the start stops for all routes have the same departure schedule [8 : 00, 8 : 15, 8 : 30, 8 : 45, 9 : 00, 14 : 00, 14 : 30, 15 : 00, 15 : 30] and the inter-stop travel time for every route is 15 min for morning trips and 10 min for afternoon trips. For example the travel time along the link that connects stop $S1$ to stop $S2$ is 15 min for start time 8:00 AM and it changes to 10 min for the start time 2:00 PM.

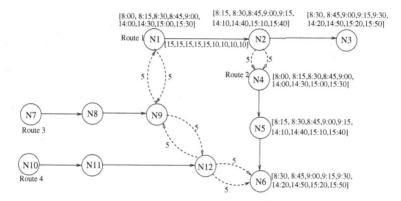

Fig. 5.2 TAG representation of a multimodal network

5.1.2 Time Aggregated Graph Representation

Figure 5.2 shows the time aggregated graph representation of the multimodal network shown in Fig. 5.1. The schedules are associated with each node and the time dependent travel times are represented along the links. The links that represent transfers across modes or routes are also associated with a cost that would include the transit time between stops. Costs of transfer links are assumed to be 5 min.

5.1.3 Routing in Multimodal Networks

The routing algorithms that are described in Chap. 3 and 4 can be used in computing least cost routes in multimodal networks. Routes can be computed based on a fixed start time or a flexible start time so that the time spent in the network is minimized.

A trace of SP-TAG algorithm from Chap. 3 is given in Table 3.3 for start node N1 and end node N6 for a start time of 8 AM. This trace assumes that the network displays the FIFO property. The table entries are the costs associated with each node (the time required to reach the node from the start) at each iteration. The node marked as "closed" is the node with the minimum cost selected for expansion.

The routing algorithm accounts for wait times wherever necessary. For example, when expanding node $N2$ the cost of node $N4$ is updated based on the cost of transfer link and the wait required before the next earliest departure time (8:30), thus updating the cost as Cost(N4) = Cost(N3)+Transfer link cost+Wait = 15 + 5 + 10 = 30.

A multimodal transportation network can violate the FIFO property. There might be cases where total travel time can be reduced by choosing to wait at a stop rather than choose the earliest departure. If the start time is fixed, NF-SP-TAG algorithm from Chap. 3 can compute the shortest path, whereas CP-NF-BEST algorithm (Chap. 4)

Table 5.1 Trace of the SP-TAG algorithm for the network shown in Fig. 5.2

Iteration	N1	N2	N3	N4	N5	N6	N9
1	0 (closed)	∞	∞	∞	∞	∞	∞
2	0	15	∞	∞	∞	∞	5 (closed)
3	0	15 (closed)	∞	∞	∞	∞	5
4	0	15	30	30 (closed)	∞	∞	5
5	0	15	30	30	45	∞	5
6	0	15	30 (closed)	30	45	∞	5
7	0	15	30	30	45	∞	5
8	0	15	30	30	45 (closed)	∞	5
9	0	15	30	30	45	60	5

could be used to compute the best time to start a journey so that the time spent in the network is minimized.

5.2 Modeling Sensor Networks

Finding novel and interesting spatio-temporal patterns in the ever increasing collection of sensor data is an important problem in several scientific domains. Many of these scientific domains collect sensor data in outdoor environments with underlying physical interactions. For example, in environmental science, a timely response to anticipated watershed/in-plant events (e.g., chemical spill, terrorism, etc.) to maintain water quality is required.

A collection of sensors may be represented as a sensor graph where the nodes represent the sensors and the edges represent selected relationships. For example, sensors upstream and downstream in a river may have physical interactions via water flow and related phenomenon such as plume propagation. Relationships can also be geographical in nature, such as proximity between the sensor units. Formulation of a model to represent a sensor graph that supports mining useful information from data poses some significant challenges. Since the volume of data is large, the model used to represent the sensor graph must be storage efficient. It should also provide sufficient support for the design of correct and efficient algorithms for data analysis. Second, the sensor graph characteristics modeled as pairs, $< measured\_value, error >$, can be time-dependent (e.g., the flow rate in a river stream). The model used to represent a time-dependent graph should be able to represent the time-variance, simultaneously maintaining the storage efficiency.

A sensor graph is spatio-temporal in nature since the relative locations of the sensor nodes and the time-dependence of their characteristics are significant. Spatio-Temporal graphs can be modeled as time expanded graphs, where the entire network is replicated for every time instant [26]. The changes in the graph can be very frequent and for modeling such frequent changes, the time expanded networks would require a large number of copies of the original network, thus leading to network sizes that are too memory expensive. Moreover, while modeling sensor graphs that involve no physical flow, a direct application of this model might not be possible.

Time aggregated graph (TAG) can model the changes in a spatio-temporal graph by collecting the node and edge attributes into a set of time series. The model can also account for the changes in the topology of the network. The edges and nodes can disappear from the network during certain instants of time and new nodes and edges can be added. TAG keeps track of these changes through a time series attached to each node and edge that indicates their presence at various instants of time. The stochastic nature of the physical relationships between the sensors (e.g., the flow rate of the river stream that connects the sensors) is accounted for by expressing each element in the attribute time series as a pair of values (i.e., ¡measured value, error¿) [15].

5.2.1 Hotspot Detection

Definition. The problem of hot spot detection is to discover the sensor nodes that display significant differences between observed values and expected 'standard' values.

Application. In application domains such as river systems where chemical levels are constantly monitored, sensors are deployed to detect the changes. In this context, a hotspot is indicated by a sensor reporting an anomaly, which is characterized by a measured value different from the expected value. A method to discover hotspots using TAG representation of sensors is briefly described in this section. The nodes in the TAG represent the sensors. An edge is added between the nodes if and only if there is a physical relationship between the nodes. The presence of a hotspot at a node at various time instants is indicated by a node time series. In addition, the time dependence of the physical relationships modeled by the edges can be represented by edge time series attributes. Figure 5.3 illustrates the graph model for the sensor graph. For the sake of simplicity, edge attributes are not shown in Fig. 5.3. Figure 5.3a shows an example network. The nodes that are active at time instants $t = 2$ and 3 are shown in Fig. 5.3b and c. The TAG representation is shown in Fig. 5.3c. The time series attributes on the nodes indicate the hotspots at various time instants. For example, the time series 2, 3 on the node N2 indicates that the node is a hot spot at $t = 2, 3$.

Method. Given a sensor graph called the source node, the hot spot at any time instant is the set of nodes where an anomaly has been detected the given time instant. A modified breadth first strategy is used to find the nodes that indicate the hot spots at any time instant. The pseudo-code is provided in Algorithm 1 [15].

Each node has a time series attribute that encodes the information about the time instants at which the node has an anomaly. For example, the time series [2, 3] at node N2 in Fig. 5.3d indicates that the node is a hotspot at $t = 2, 3$. The algorithm searches the graph starting at (any) given node for each value of time t and finds the

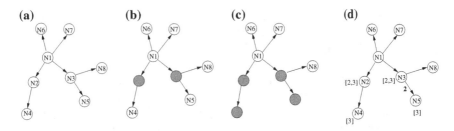

Fig. 5.3 TAG model to detect hotspots

Table 5.2 Execution trace of the hotspot algorithm

time	t1	t2	t3
Hotspot nodes	∅	{N2, N3}	{N2, N3, N4, N5}

hotspots. The search uses a breadth-first strategy, modified to incorporate the fact that each node has a time series that needs to be checked. When each node is visited, the algorithm checks to see whether it is a hot spot by checking the node time series. The node time series is assumed to be sorted. The output of the algorithm is the set of hotspots at every time instant.

Algorithm 1: Hotspot Algorithm

1: Function BASICHOTSPOTS(Graph $G(N, E)$, set N, set E, node $source$)
2: **for** $t = 1, T$ **do**
3:     mark $source$ as visited;
4:     enqueue(Q,source);
5:     **if** t in node_time_series of source **then**
6:         hotspots[t] = source;
7:     **end if**
8:     **while** Q not empty **do**
9:         u = Dequeue();
10:        For every node v such that $uv \in E$ and if $exists(nodeu, t)$
11:            **if** v is not marked **then**
12:                $v = visited$
13:                $Enqueue(Q, v)$;
14:                $hotspots[t] = hotspots[t] \bigcup v$;
15:            **end if**
16:        **end while**
17: **end for**

Execution Trace. Table 5.2 shows the trace of the algorithm for the network shown in Fig. 5.3d. The search starts at node N1 at $t = 1$ and detects no hotspots. At $t = 2$, the search finds that the nodes N2 and N3 are hotspots based on the entry '2' (indicating the presence of a hotspot at $t = 2$) in their node time series [2, 3]. The algorithm performs another iteration for $t = 3$ and finds the hotspots at N2, N3, N4, and N5. The execution trace is summarized in Table 5.2.

References

1. R. Ahuja, T. Magnanti, J. Orlin, *Network Flows—Theory, Algorithms, and Applications*, (Prentice Hall, Englewood Cliffs, 1993)
2. D. Bertsekas, *Dynamic Programming: Deterministic and Stochastic Models* (Prentice Hall, Englewood Cliffs, 1987)
3. P. Chen, The entity-relationship model—towards a unified view of data. ACM Trans. Database Syst. **1**(1), 9–36 (1976)
4. B. Cherkassky, A. Goldberg, T. Radzik, Shortest paths algorithms: theory and experimental evaluation. Math. Program. **73**, 129–174 (1996)
5. T. Cormen, C.E. Leiserson, R.L. Rivest, C. Stein, *Introduction to Algorithms (Chapter 26, Flow Networks)* (MIT Press, Cambridge, 2002)
6. O. Corporation. Oracle spatial and oracle locator: location features for oracle. http://www.oracle.com/technology/products/spatial/
7. O. Corporation. Oracle spatial 10g: an oracle white paper. http://www.oracle.com/technology/products/spatial/ 2005
8. B.C. Dean, Algorithms for minimum-cost paths in time-dependent networks. Networks **44**(1), 41–46 (2004)
9. C.H. Deutsch, U.P.S. embraces high-tech delivery methods. The New York Times (http://www.nyt.com/) (July 12 2007)
10. Z. Ding, R. Guting, Modeling temporally variable transportation networks, in *Procceedings 16th International conference on Database Systems for Advanced Applications* (Houston, 2004), pp. 154–168
11. M. Erwig, Graphs in Spatial Databases. Ph.D. thesis, Fern Universität Hagen, 1994
12. M. Erwig, R. Guting, Explicit graphs in a functional model for spatial databases. IEEE Trans. Knowl. Data Eng. **6**(5), 787–804 (1994)
13. ESRI. ArcGIS network analyst. http://www.esri.com/software/arcgis/extensions/ 2006
14. L. Ford, D. Fulkerson, Constructing maximal dynamic flows from static flows. Oper. Res. **6**, 419–433 (1958)
15. B. George, J. Kang, S. Shekhar, STSG: a data model for the discovery of spatio-temporal patterns, in *Proceedings of First International Workshop on Knowledge Discovery from Sensor Data in conjunction with ACM SIGKDD International Conference on Knowledge Discovery and Data Mining (KDD 2007)*, Aug 2007
16. B. George, S. Shekhar, Time-aggregated graphs for modeling spatio-temporal networks—an extended abstract, in: *Proceedings of Workshops at International Conference on Conceptual Modeling*, Nov 2006

B. George and S. Kim, *Spatio-Temporal Networks*,
SpringerBriefs in Computer Science, DOI: 10.1007/978-1-4614-4918-8,
© The Author(s) 2013

17. B. George, S. Shekhar, Time aggregated graphs: a model for spatio-temporal network. J. Data Semant. 1(2), 249–303 (2007)
18. A. Goldberg, Network optimization library. http://www.avglab.com/andrew/soft.html 2002
19. H. Gregerson, C. Jensen, Temporal entity relationship models—a survey. IEEE Trans. Knowl. Data Eng. 11(3), 464–497 1999
20. R. Hall, The fastest path through a network with random time-dependent travel times. Transp. Sci. 20, 182–188 (1986)
21. R. Hall (ed.), Handbook of Transportation Science (Kluwer Academic Publishers, Norwell, 2003)
22. H. Hamacher, S. Tjandra, Mathematical Modeling of Evacuation Problems: A Sate of the Art. Pedestrian and Evacuation Dynamics, (Springer, Berlin, 2002) pp. 227–266
23. T. Hamre, Development of semantic spatio-temporal data models for integration of remote sensing and in situ data in marine information system. Ph.D. thesis, University of Bergen, 1995
24. P. Hart, N. Nilsson, B. Raphael, A formal basis for the heuristic determination of minimum cost paths. IEEE Trans. Syst. Sci. Cybern. 4(2), 100–107 (1986)
25. D. Kaufman, R. Smith, Fastest paths in time-dependent networks for intelligent vehicle highway systems applications. IVHS J. 1(1), 1–11 (1993)
26. E. Kohler, K. Langtau, M. Skutella, Time-expanded graphs for flow-dependent transit times, in Procceedings 10th Annual European Symposium on Algorithms, pp. 599–611, 2002
27. M. Koubarakis, T.K. Sellis, A.U. Frank, S. Grumbach, R.H. Güting, C.S. Jensen, N.A. Lorentzos, Y. Manolopoulos, E. Nardelli, B. Pernici, H.-J. Schek, M. Scholl, B. Theodoulidis, N. Tryfona, (eds.), Spatio-Temporal Databases: The CHOROCHRONOS Approach, Lecture Notes in Computer Science, vol. 2520, (Springer, Berlin, 2003)
28. N. Levine, CrimeStat 3.0: A Spatial Statistics Program for the Analysis of Crime Incident Locations. (Ned Levine& Associatiates, Houston, 2004)
29. D. Liu, S. Shekhar, M. Coyle, S. Sarkar. An evaluation of access methods for spatial netwroks in Proceedings of the 2nd workshop on Advances in Geographic, Information Systems, 1994
30. Q. Lu, B. George, S. Shekhar, Capacity constrained routing algorithms for evacuation planning: a summary of results, in Proceedings of 9th International Symposium on Spatial and Temporal Databases (SSTD'05), Aug 2005
31. E. Miller-Hooks, H. Mahmassani, Least possible time paths in stochastic time-varying networks. Comput. Oper. Res. 25(12), 1107–1125 (1998)
32. E. Miller-Hooks, H. Mahmassani, Path comparisons for a priori and time-adaptive decisions in stochastic, time-varying networks. Eur. J. Oper. Res. 146, 67–82 (2003)
33. Oracle. Oracle spatial 10g, an oracle white paper. http://www.oracle.com/technology/products/spatial/, August 2005
34. A. Orda, R. Rom, Minimum weight paths in time-dependent networks. Networks 21, 295–319 (1991)
35. S. Pallottino, Shortest-path methods: complexity, interrelations and new propositions. Networks 14, 257–267 (1984)
36. S. Pallottino, M.G. Scuttella, Equilibrium and Advanced Transportation Modelling . Shortest Path Algorithms in Tranportation Models: Classical and Innovative Aspects. (Kluwer, Boston, 1998), pp. 245–281
37. J. Pearl, Heuristics: Intelligent Search Strategies for Computer Problem Solving, (Addison Wesley, Reading, 1984)
38. J. Rasinmäki, Modelling spatio-temporal environmental data in 5th AGILE Conference on Geographic Information Science, Palma, 2002
39. S. Russel, P. Norwig, Artificial Intelligence: A Modern Approach (Prentice Hall, Upper Saddle River, 1995)
40. S. Shekhar, S. Chawla, Spatial Databases: Tour (Prentice Hall, Englewood- Cliffs, 2003)
41. D. Sawitzki, Implicit Maximization of Flows over Time. Technical report, University of Dortmund, 2004

42. S.E. Dreyfus, An appraisal of some shortest path algorithms. Oper. Res. **17**, 395–412, (1969)
43. Y. Sheffi, *Urban Transportation Networks: Equilibrium Analysis with Mathematical Programming Method* (Prentice-Hall, Englewood Cliffs, 1985)
44. S. Shekhar, D. Liu, Connectivity-clustered access method for networks and networks computations: Summary of results, in *Proceedings of IEEE International Conference on Data Engineering*, 1995
45. S. Shekhar, D. Liu, CCAM: A connectivity-clustered access method for networks and networks computations. IEEE Trans. Knowl. Data Eng. (1997)
46. S. Shekhar, R. Vatsavai, S.Chawla, T. Burk, Spatial Pictogram Enhanced Conceptual Data Models and Their Translation to Logical Data Models. *Integrated Spatial Databases: Digital Images and GIS., Lecture Notes in Computer Science*, Vol. 1737, ed. by P. Agouris, A. Stefanidis (Springer, Berlin, 1999)
47. S. Stephens, J. Rung, X. Lopez, Graph data representation in oracle databese 10g: case studies in life sciences. IEEE Data Eng. Bull. **27**(4), 61–66 (2004)
48. J. Wardrop, Some theoretical aspects of road traffic research, in *Proceedings of the Institution of Civil Engineers*, **2**(1), 1952
49. F. Zhan, C. Noon, Shortest paths algorithms: an evaluation using real road networks. Transp. Sci. **32**, 65–73 (1998)
50. E. Zimayi, C. Parent, S. Spaccapietra, TERC+: a temporal conceptual model, in *Proceedings of International Symposium on Digital Media Information Base* Nov 1997